AUSTRALIAN MATHEMATICAL SOCIETY LECTURE SERIES

1 Introduction to Linear and Convex Programming, N. CAMERON
2 Manifolds and Mechanics, A. JONES, A. GRAY & R. HUTTON

Australian Mathematical Society Lecture Series. 2

Manifolds and Mechanics

Arthur Jones and Alistair Gray
Mathematics Department, La Trobe University

Robert Hutton
Comalco Ltd

CAMBRIDGE UNIVERSITY PRESS

Cambridge

New York New Rochelle

Melbourne Sydney

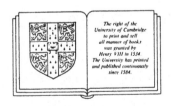

CAMBRIDGE UNIVERSITY PRESS
Cambridge, New York, Melbourne, Madrid, Cape Town, Singapore, São Paulo, Delhi

Cambridge University Press
The Edinburgh Building, Cambridge CB2 8RU, UK

Published in the United States of America by Cambridge University Press, New York

www.cambridge.org
Information on this title: www.cambridge.org/9780521333757

First published 1987
Reprinted 1988
Re-issued in this digitally printed version 2008

A catalogue record for this publication is available from the British Library

ISBN 978-0-521-33375-7 hardback
ISBN 978-0-521-33650-5 paperback

CONTENTS

PROLOGUE

These notes began as an attempt to bridge the gap between undergraduate advanced calculus texts (such as Spivak(1965)) and graduate texts on differential topology. One of the major applications of differentiable manifolds is to the foundations of mechanics. Here a huge gap exists between the classical literature (see for example Goldstein (1980) and the more modern differentiable manifolds approach (see Abraham and Marsden(1978)). Dieudonne(1972) says "The traditional domain of differential geometry, namely the study of curves and surfaces in three-dimensional space, was soon realized to be inadequate, particularly under the influence of mechanics". Two of the authors of this work have been involved in the writing of papers on problems in Lagrangian mechanics and they soon realized the need for a simple but modern treatment of the theoretical background to such problems.

Thus our aim is to make some of the basic ideas about manifolds readily available to applied mathematicians and theoretical physicists while at the same time exhibiting applications of an important area of modern mathematics to mathematicians.

Classical texts use vague ideas such as "virtual work" and "infinitesimal displacements" in their derivation of Lagrange's equations. By contrast the modern texts jump straight to Hamiltonian systems and lose the physical motivation. In these notes, the derivation of Lagrange's equations makes direct appeal to geometrical principles. As far as the authors are aware, these ideas do not appear elsewhere.

In his original work, *Mécanique analytique*, Lagrange used no diagrams. In fact the geometrical ideas necessary for a thorough understanding of classical mechanics did not exist at that time. It was only with the development of differential geometry and differential topology that geometry once again became an essential part of classical mechanics.

An advantage of starting with the Lagrangian formulation, as opposed to the Hamiltonian, is that it follows directly from Newton's Laws of Motion and that the mathematical background required is much less formidable.

These notes were developed primarily with a view to providing the necessary theory for the study of particle motion as in the papers by Gray *et al*(1982) and Gray(1983). As we demonstrate, however, our treatment of Lagrangian mechanics is equally applicable to the motion of rigid bodies which consist of a finite collection of particles.

Our approach can be extended so as to apply to the motion of a rigid body defined by a continuous mass distribution even in nonholonomic cases but this work is beyond the scope of the present volume and will appear separately.

We thank Lindley Scott for the diagrams of page 132, Sid Morris for his encouragement, the referees for their constructive reports and Professor Sneddon for reading and commenting on our preliminary notes.

Our debt to Judy Storey is profound and we thank her accordingly. Without her monumental effort this manuscript would never have reached its final form.

Arthur Jones
Alistair Gray
Robert Hutton
December, 1986.

1. CALCULUS PRELIMINARIES

The differential calculus for functions which map one normed vector space into another normed vector space is the main prerequisite for the study of manifolds and Lagrangian mechanics in the modern idiom. The mild degree of abstraction involved helps focus attention on the simple geometric ideas underlying the basic concepts and results. The insights and techniques which this approach fosters turn out to be very worthwhile in the study of many topics in applied mathematics.

The idea of a function as an entity in its own right, independently of any numerical variables which may be used to define it, is fundamental for this approach to calculus. This idea, although usually confined to statements of the theory, can easily be used to provide computational tools for particular examples. This involves the development of a "variable free" language of calculus in which the functions themselves, as distinct from the values that they take are highlighted. This enables us to formulate and work with many important and difficult mathematical ideas in a simple manner. Some of the notation does not appear elsewhere but a good account of the basic ideas may be found in Spivak(1965) or in Lang(1964).

1.1. FRECHET DERIVATIVES

Here we outline how the idea of a derivative can be formulated for functions which map one normed vector space to another. The basic idea is this: a function is differentiable at a point if there is an affine map which approximates the function very closely. The derivative of the function at this point is then the linear part of this affine map.

1.1.1. Example. Let $f\colon R^2 \to R^2$ be given by

$$f(x,y) = (x^2 + y^2 + 1,\ xy + 1)$$

so that

$$f(1+h,\ 2+k) = f(1,2) + (2h+4k,\ 2h+k) + (h^2+k^2,\ hk).$$

As (h,k) approaches $(0,0)$ the quadratic terms become negligible compared with h and k and so f is approximated very closely by the affine map $A\colon R^2 \to R^2$ with

$$A(1+h,\ 2+k) = f(1,2) + (2h+4k,\ 2h+k).$$

The linear part of A is the linear map $L\colon R^2 \to R^2$ with

$$L(h,k) = (2h+4k,\ 2h+k).$$

The precise meaning of "approximates f closely" is clarified in the following definition by the use of limits.

1.1.2. Definition. Let $f\colon U \longrightarrow F$ where E and F are normed vector spaces and U is an open subset of E. By saying that f is *differentiable* at $a \in U$ we mean that there is a continuous affine map $A\colon E \longrightarrow F$ such that

$$f(a) = A(a)$$

and

$$\lim_{x \to a} \frac{\|f(x)-A(x)\|}{\|x-a\|} = 0. \quad \blacksquare$$

1.1.3. Definition. For a map f as in Definition 1.1.1 we define its *Fréchet derivative* at a to be the unique continuous linear map $Df(a)\colon E \longrightarrow F$ such that

$$\lim_{h \to 0} \frac{\|f(a+h)-f(a)-Df(a)(h)\|}{\|h\|} = 0. \quad \blacksquare$$

Now let us return to Example 1.1.2. and verify that in that case

$$Df(1,2)(h,k) = (2h+4k,\ 2h+k).$$

Notice that the norm of a vector $(x,y) \in R^2$ is given by $\|(x,y)\| = \sqrt{x^2+y^2}$ and so in this case we have

$$\lim_{(h,k)\to(0,0)} \frac{\|f(1+h,2+k) - f(1,2) - (2h+4k,2h+k)\|}{\|(h,k)\|}$$

$$= \lim_{(h,k)\to(0,0)} \frac{\|(h^2+k^2,\ hk)\|}{\|(h,k)\|}$$

and since the terms in the numerator are quadratic it is easy to show that this limit is in fact zero.

Notice that the Jacobian matrix for f at the point $(1,2)$ is just the matrix of the linear map $Df(1,2)$. More about this later.

1.1.4. Theorem. (Chain Rule I). *Let maps* $f\colon U \to F$ *and* $g\colon F \to G$ *be differentiable at points* $a \in U$ *and* $f(a) \in F$ *respectively where* E, F *and* G *are normed vector spaces and* U *is an open subset of* E. *The composite map* $g \circ f\colon U \to G$ *is then differentiable at* a *and*

$$D(g \circ f)(a) = Dg(f(a)) \circ Df(a).$$

A proof of this theorem may be found in Spivak(1965). ■

A variable-free notation for functions which map normed vector spaces to normed vector spaces (in most applications to mechanics these normed vector spaces are, or are closely related to, R^n) turns out to be very useful for our purposes.

1.1.5. Definition. The *identity mapping from a normed vector space* E *to itself* is denoted by id_E and defined by

$$id_E(x) = x \quad \text{for each } x \in E. \quad ■$$

1.1.6. Definition. The *constant mapping from* E *to* F (E, F normed vector spaces) which takes the value $c \in F$ is denoted by $\underline{c}$ and defined by

$$\underline{c}(x) = c \quad \text{for each } x \in E. \quad ■$$

Usually we will omit the subscript "E" from id_E when the context is clear.

EXERCISES 1.1.

1. *Show that* $D\, id_E(a) = id_E$ *for each* $a \in E$.

2. *Show that if* f *is a continuous linear map between normed vector spaces then* $Df(a) = f$.

3. *Suppose* $\phi\colon \mathsf{R}^n \to \mathsf{R}^n$ *is differentiable and has a differentiable inverse (that is,* ϕ *is a diffeomorphism). Show that* $D\phi^{-1}(\phi(a)) = (D\phi(a))^{-1}$ *for each* $a \in \mathsf{R}^n$.

4. *Let* M^n *denote the set of all real* $n \times n$ *matrices. It can be shown that* M^n *can be made into a normed vector space in such a way that for all* $A, B \in M^n$

$$\|AB\| \leqslant \|A\|\,\|B\| .$$

Now define $f: M^n \to M^n$ *by*

$$f(X) = X^2 .$$

Show that f *is differentiable at* $A \in M^n$ *and find its Fréchet derivative* $Df(A)$.

1.2. THE TANGENT FUNCTOR

The ideas of this section will later be generalized to sets which are a little more involved than normed vector spaces.

1.2.1. Definition. For an open set U in the normed vector space E we define the *tangent space* T_aE at $a \in U$ as the set
$T_aE = \{(a,h): h \in E\}$. ■

This tangent space can easily and naturally be given a vector space structure.

1.2.2. Definition. *Vector addition and scalar multiplication on* T_aE are defined by putting

$$(a,h) + (a,k) = (a, h+k)$$
$$\lambda(a,h) = (a,\lambda h)$$

for all vectors (a,h) and $(a,k) \in T_aE$ and scalars $\lambda \in R$. ■

We may think of a typical vector $(a,h) \in T_aE$ as an arrow emanating from a. Its tail is at a and its tip at $a + h$. See Figure 1.2.1.

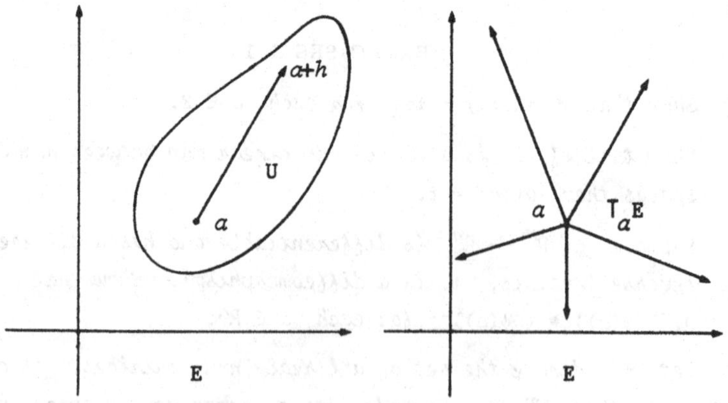

Figure 1.2.1. Tangent Space at a.

1.2.3. Definition. The *tangent bundle* of U at a is defined as

$$\bigcup_{a \in U} T_a E = U \times E. \blacksquare$$

It is clear from Section 1.1 that the derivative of a map $f: U \subset E \to F$ (E, F normed vector spaces) U open, at a point $a \in U$ is the linear map

$$Df(a) \in \mathcal{L}(E,F)$$

and that in some sense

$$A(x) = f(a) + Df(a)(x-a)$$

is an "affine approximation" to $f(x)$ near a. (See Figure 1.2.2). The two important parts of this idea of affine approximation are the linear map $Df(a)$ and the point $f(a)$. In the usual formulation of differentiation the point $f(a)$ is "lost" and the derivative $Df(a)$ is all that is retained.

It turns out that the tangent map overcomes this problem (and in doing so, leads to a rather neat version of the chain rule). The idea is that whereas

the derivative $Df(a)$ *maps* E *to* F,
the tangent map $T_a f$ should map $T_a E$ to $T_{f(a)}F$.

Notice that, as in Figure 1.2.3, we can think of the vectors h and $Df(a)(h)$ as being "based at" a and $f(a)$ respectively.

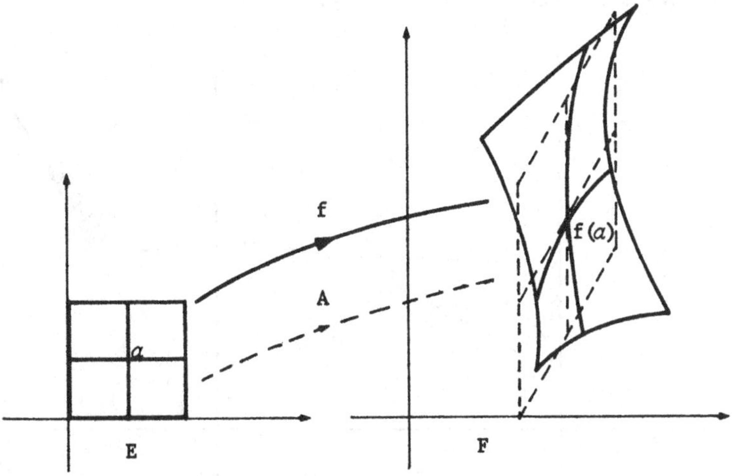

Figure 1.2.2. Affine Approximation at a.

1.2.4. Definition. Let $f\colon U \to F$ be differentiable at $a \in U$ where $U \subseteq E$ is open (E, F normed vector spaces). Then *the tangent map at* a *of* f is the map

$$T_a f\colon T_a E \to T_{f(a)} F$$

defined by

$$T_a f(a,h) = (f(a), Df(a)(h)) \text{ for each } (a,h) \in T_a E. \quad \blacksquare$$

It is clear that the tangent map is linear. Now by allowing the point a to vary throughout U we obtain a map defined on the whole of the tangent bundle.

1.2.5. Definition. Let f, U, E, F, a be as in Definition 1.2.4. Then the *tangent functor*

$$Tf\colon U \times E \to F \times F$$

is defined by

$$Tf(a,h) = (f(a), Df(a)(h)). \quad \blacksquare$$

We are now in a position to formulate a very neat version of the chain rule. You may find the comparison between the theorem below and Theorem 1.1.4 interesting.

1.2.6. Theorem. (Chain Rule II) *Suppose* $g\colon U \subset E \to V \subset F$ *and* $f\colon V \to G$ *where* E, F, G *are n.v.s. and* U, V *are open and* f, g *are differentiable. Then* $T(f \circ g) = Tf \circ Tg$.

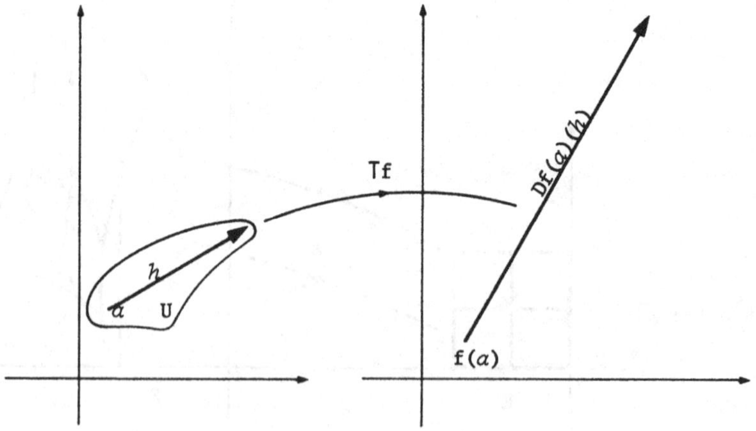

Figure 1.2.3. The tangent map at a.

Proof. Let $a \in U$, $h \in E$. Then

$$\begin{aligned} T(f\circ g)(a,h) &= (f\circ g(a),\ D(f\circ g)(a)(h)) \\ &= (f(g(a)),\ D\,f(g(a))\circ Dg(a)(h)) \text{ by Theorem 1.1.4.} \\ &= Tf(g(a),\ Dg(a)(h)) \\ &= Tf(Tg(a,h)). \end{aligned}$$

Thus $T(f\circ g) = Tf\circ Tg$ as required. ■

This property of T, as given in Theorem 1.2.6, is usually expressed by saying that T is a *functor*.

EXERCISES 1.2

1. *Check that*
   (a) $T\,id_E = id_{E\times E}$
   (b) $Tf^{-1} = (Tf)^{-1}$

   (provided of course f^{-1} *exists and is differentiable.)*

2. *Suppose* $\Pi : E \times E \to E : (a,b) \mapsto a$ *and, by abuse of notation*

$$\Pi : F \times F \to F : (x,y) \mapsto x$$

Show that the diagram in Figure 1.2.4 commutes.

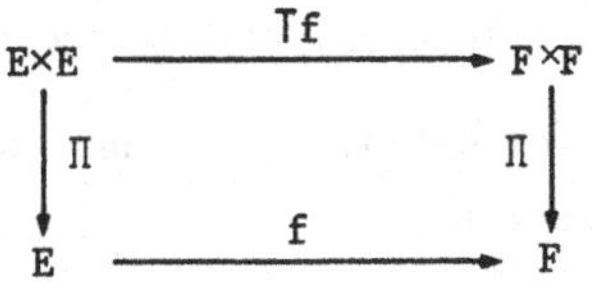

Figure 1.2.4.

3. *Let* $f: (0,2\pi) \to R^2$ *be given by* $f(a) = (\cos(a),\sin(a))$.
   (a) *Sketch the image under* f *of* $(0, 2\pi)$.
   (b) *Compute* $Tf(a,1)$.
   (c) *Sketch* $Tf(a,1)$ *as an element of* $T_{f(a)}R^2$.

1.3 PARTIAL DIFFERENTIATION

The idea of a Fréchet derivative is readily generalized to include partial derivatives.

1.3.1. Definition. Let E_1, E_2 and F be normed vector spaces, U_1, U_2 be open subsets of E_1 and E_2 respectively and suppose $f: U_1 \times U_2 \to F$ is a differentiable function. Let $(a,b) \in U_1 \times U_2$.

The partial derivative of f *with respect to its first variable,* $\partial_1 f(a,b)$, is the unique continuous linear map from E_1 to F such that

$$\lim_{h \to 0} \frac{\|f(a+h,b) - f(a,b) - \partial_1 f(a,b)(h)\|}{\|h\|} = 0. \quad \blacksquare$$

The partial derivative $\partial_2 f(a,b)$ may be defined in a similar way. A version of the chain rule for partial derivatives is then given by the following Theorem.

1.3.2. Theorem. (Chain Rule for Partial Derivatives I).

Let E_1, E_2, E *and* F *be normed vector spaces,* U_1 *and* U_2 *be open subsets of* E_1 and E_2 *respectively*

$$f: U_1 \times U_2 \to E$$

be differentiable at $(a,b) \in U_1 \times U_2$ *and*

$$g: E \to F$$

be differentiable at $f(a,b) \in E$. *Then for* $i = 1,2$

$$\partial_i(g \circ f)(a,b) = Dg(f(a,b)) \circ \partial_i f(a,b). \quad \blacksquare$$

It is easy to see how Definition 1.3.1. and Theorem 1.3.2. can be generalized to a "function of n variables".

1.3.3. Lemma. *Let* $f: \mathbb{R}^n \to \mathbb{R}$ *be differentiable at* $a \in \mathbb{R}^n$. *Then*

$$Df(a)\left(\sum_{j=1}^{n} x_j e_j\right) = \sum_{j=1}^{n} \partial_j f(a)(x_j)$$

where the e_j *denote the usual basis vectors for* $\mathbb{R}^n$. $\blacksquare$

We will use id_j to denote the projection function

$$id_j: \mathbb{R}^n \to \mathbb{R}: (x_1,\ldots,x_n) \mapsto x_j .$$

EXERCISES 1.3

1. *Use the chain rule for partial derivatives to show that if* $f\colon R^2 \to R^2$ *is differentiable then* $\partial_i(\mathrm{id}_j \circ f) = \mathrm{id}_j \circ \partial_i f$ *for each* $i,j = 1,2$.

2. *Let* $\mathbf{C}^1([0,1])$ *denote the set of once differentiable functions on the domain* $[0,1]$. *We define the evaluation map*

$$\mathrm{ev}\colon \mathbf{C}^1([0,1]) \times [0,1] \to R\colon (f,x) \to f(x).$$

Find $\partial_1 \mathrm{ev}(f,x)(g)$

and $\partial_2 \mathrm{ev}(f,x)(y)$.

1.4. COMPONENTWISE CALCULUS

Many of the important results in mechanics rely on the chain rule in its various forms. When we deal with coordinates, which we will later call "charts", a componentwise but variable free version of the chain rule proves to be extremely useful. Thus in this section we concentrate on derivatives of maps between R^n and R^m.

We will begin by introducing some (nonstandard) notation for partial derivatives of real valued functions.

1.4.1. Definition. Let $f\colon R^n \to R$ be differentiable at $a \in R^n$. Then the *slope of* f *with respect to its ith variable* $(i = 1,2,\ldots,n)$ is denoted by $f^{/i}$ and defined by $f^{/i}(a) = \partial_i f(a)(1)$. ■

The advantages of this notation will become apparent when we are faced with the task of formulating a readable version of the chain rule for maps between R^n and R^m. Note that classical texts would write

$$\left.\frac{\partial f}{\partial x_i}\right|_{x=a} \quad \text{for} \quad f^{/i}(a) \qquad \text{and that}$$

$f^{/i}(a)$ is a real number as distinct from $\partial_i f(a)$ which is a linear map. A further link with the classical treatment is provided by the following definition and theorem.

1.4.2. Definition. Let $f\colon R^n \to R^m$ be differentiable. Then the matrix of the derivative of f at $a \in R^n$, that is the matrix of $Df(a)$, is called the *Jacobian matrix of* f *at* a and is denoted by $f'(a)$. ■

1.4.3. Theorem. *Let* $f: \mathbb{R}^n \to \mathbb{R}^m$ *be differentiable. Then the ij-th component of the Jacobian matrix of* f *at* a *is given by* $f_i^{/j}(a)$.

Proof. Denote f by the ordered m-tuple of its component functions, that is,

$$f = (f_1, f_2, \ldots, f_m)$$

and so $f_i : \mathbb{R}^n \to \mathbb{R}$ for each $i = 1,2,\ldots,m$. Application of Lemma 1.3.3. to each of the f_i then gives the required result. ∎

1.4.4. Corollary. *Let* $f: \mathbb{R} \to \mathbb{R}^m$ *be differentiable. Then for each* $x \in \mathbb{R}$

$$Df(x)(1) = f'(x) = \lim_{h \to 0} \frac{f(x+h) - f(x)}{h}.$$

Proof. Follows from Theorem 1.4.2. and Definitions 1.4.1. and 1.3.1. ∎

We are now in a position to formulate a "component" version of Theorem 1.3.2.

1.4.5. Theorem. (Chain Rule for Partial Derivatives II).
Let $h: \mathbb{R}^n \to \mathbb{R}^p$, $g: \mathbb{R}^m \to \mathbb{R}^n$ *be differentiable. Then for each* $i = 1,2,\ldots,m$

$$(h \circ g)^{/i} = \sum_{j=1}^{n} h^{/j} \circ g \; g_j^{/i}.$$

Proof. Let $a \in \mathbb{R}^m$. Then for each $i = 1,2,\ldots,m$

$$\begin{aligned} (h \circ g)^{/i}(a) &= \partial_i (h \circ g)(a)(1) && \text{(Definition 1.5.1.)} \\ &= Dh(g(a)) \circ \partial_i g(a)(1) && \text{(Theorem 1.3.2.)} \\ &= Dh(g(a))(g^{/i}(a)) && \text{(Definition 1.5.1.)} \end{aligned}$$

Now note that $g^{/i}(a) = (g_1^{/i}(a), g_2^{/i}(a), \ldots, g_n^{/i}(a))$ and an application of Theorem 1.4.3. is sufficient to complete the proof. ∎

EXERCISES 1.4

1. *Refer to example 1.1.1. Verify Theorem 1.4.3. for this case.*

2. *Fill in the details of the proofs of Theorems 1.4.3. and 1.4.5. and Corollary 1.4.4.*

3. *Let* $\phi: \mathbb{R}^m \to \mathbb{R}^n$, $\gamma: \mathbb{R} \to \mathbb{R}^m$ *be differentiable. Show that*

$$(\phi \circ \gamma)'(t) = D\phi(\gamma(t))(\gamma'(t))$$

for each $t \in \mathsf{R}$.
(Hint: use Corollary 1.4.4. and Theorem 1.1.4.).

1.5. VARIABLE-FREE ELEMENTARY CALCULUS

The use of a variable-free notation in elementary calculus provides a means by which many proofs can be greatly simplified and many ambiguities removed. The ideas presented here will be used to some effect in later chapters.

The identity and constant mappings as defined in Section 1.1. here become

$$\mathrm{id}: \mathsf{R} \to \mathsf{R}: x \mapsto x \quad \text{and}$$
$$\underline{c}: \mathsf{R} \to \mathsf{R}: x \mapsto c.$$

Thus for example $\mathrm{id}^3(x) = x^3$, and $(\mathrm{id}^{-3} + \underline{3})(x) = \frac{1}{x^3} + 3$ for each $x \neq 0$. The derivative of a function $f: \mathsf{R} \to \mathsf{R}$ will be denoted by f'. For example $\cos' = -\sin$ and $(\mathrm{id}^3)' = 3\mathrm{id}^2$. We are now in a position to formulate some of the basic ideas of elementary calculus in this language.

1.5.1. Lemma. (Chain Rule). *For* $f: \mathsf{R} \to \mathsf{R}$, $g: \mathsf{R} \to \mathsf{R}$ *where* f *and* g *are differentiable*

$$(f \circ g)' = (f' \circ g) \cdot g' .$$

Proof. See any standard calculus textbook. ∎

1.5.2. Example. $(\cos \circ \mathrm{id}^3)' = -\sin \circ \mathrm{id}^3 . (3\mathrm{id}^2)$
$= -3\mathrm{id}^2 . \sin \circ \mathrm{id}^3$. ∎

1.5.3. Definition. The *indefinite integral from* a of an integrable function $f: \mathsf{R} \to \mathsf{R}$ is denoted by $\int_a f$ for $a \in \mathsf{R}$ and is defined by $\left(\int_a f\right)(x) = \int_a^x f$ for each $x \in \mathsf{R}$. ∎

1.5.4. Theorem. (Fundamental Theorem of Calculus).

(i) *Let* $f: I \to \mathsf{R}$ *be continuous, where* I *is an interval. Then for each* $a \in I$

$$\left[\int_a f\right]' = f.$$

(ii) *Let* $\phi': I \to \mathsf{R}$ *be continuous, where* I *is an interval. Then, for each* $x \in I$ *and* $a \in I$

$$\int_a^x \phi' = \phi(x) - \phi(a).$$

The proof of this theorem may be found in most respectable books on elementary calculus. ∎

An excellent example of the power and efficacy of a variable-free approach is to be found in the proof of the next theorem. This is a result that will prove useful in Chapter 12. It is instructive to compare the treatment here with others. See for example Abraham & Marsden(1978), page 63.

1.5.6. Lemma. (Gronwall's Inequality)

Let $f: [0,a] \to R^+$ *and suppose*

$$f \leqslant \underline{C} + \int_0 Kf \quad on \quad [0,a]$$

where C, K *are each positive constants.*

Then $f \leqslant Ce^{K.\mathrm{id}}$ on $[0,a]$.

Proof. First we define $g: [0,a] \to R^+$ by

$$g = \underline{C} + \int_0 Kf$$

Thus $f \leqslant g$ on $[0,a]$. Differentiating the above using Theorem 1.5.5. gives

$$g' = Kf \leqslant Kg$$

Now using the fact that $g(0) = C$ and an integrating factor $e^{-K.\mathrm{id}}$ in the differential inequation

$$g' = Kg \leqslant \underline{0}$$

gives $f \leqslant g \leqslant Ce^{K.\mathrm{id}}$ on $[0,a]$ as required. ∎

Gronwall's inequality is the cornerstone of the proof that solutions of certain differential equations vary continuously with the initial conditions.

EXERCISES 1.5

1.(a) *Use the chain rule in variable free form to find the derivative of a function* h *defined by*

$$h(x) = \arctan(x) + \arctan\left(\frac{1}{x}\right) \quad on \quad R\backslash\{0\}.$$

(Recall that $\arctan' = \dfrac{1}{\underline{1} + \mathrm{id}^2}$.*)*

(b) *Show that* h *is not a constant function by evaluating* h(1) *and* h(-1).

(c) *Use the fundamental theorem of calculus to show that if* $G' = \underline{0}$ *on an interval* I *then* G *is constant on* I.

(d) *Explain the apparent contradiction between* (b) *and* (c) *and hence sketch the function* h.

2. *Consider the initial value problem*

$$\phi' + f.\phi = g, \quad \phi(0) = \ell$$

where f,g *are continuous on* R. *Use the fundamental theorem of calculus to show that this problem is equivalent to the integral equation*

$$\phi = \underline{\ell} + \int_0 (f\phi + g).$$

(Remember to prove the implication both ways.)

2. DIFFERENTIABLE MANIFOLDS

The idea of a differentiable manifold is a combination of ideas from both analysis and geometry.

In geometry, differentiable manifolds include such things as curves and surfaces and their higher dimensional analogues. Many of the ideas to be introduced in this chapter are in fact motivated by the study of the earth's surface in elementary geography.

In analysis, the role of differentiable manifolds is to provide a natural setting, generalizing that of normed vector spaces, in which to study differentiable functions. Thus the theory developed in Chapter 1, permits you to differentiate functions mapping one normed vector space (or some open subset thereof) into another normed vector space. But once you have studied manifolds, you will also be able to differentiate functions which map, say, a sphere into a torus.

2.1. CHARTS AND ATLASES

Although the surface of the earth is a sphere, small enough regions on it will appear flat, like a plane. A geographical atlas is in fact just a collection of maps or pictures, each of which lies in a plane.

Each such picture determines a function ϕ from some region U of the earth's surface into the plane R^2 as shown in Figure 2.1.1. This idea leads to our first definition, in which we replace the earth's surface by an arbitrary set M and ignore nearly everything about the function ϕ except that it maps one-to-one onto some "flat" region —— that is, an open set in some Euclidean space R^n. The integer n will be assumed fixed for the rest of this section.

2.1.1. Definition. Let M be a set. A *chart* for M consists of a subset U of M together with a one-to-one function ϕ which maps U onto an

open set in R^n. ■

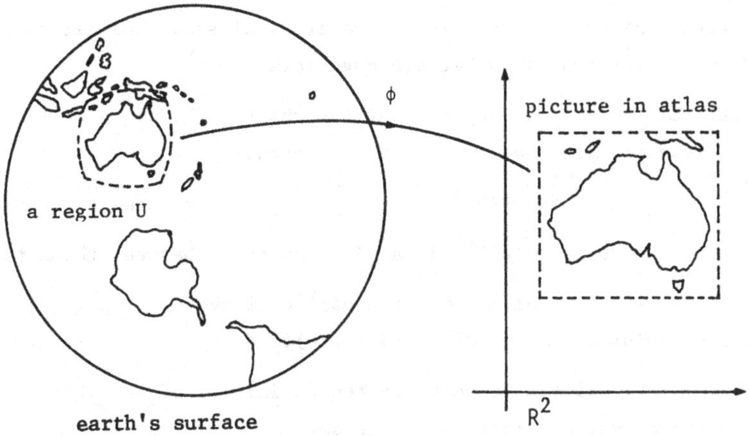

Figure 2.1.1. A chart (U,ϕ)

Before collecting together the individual charts to build up an "atlas" for the entire set M, we introduce a certain compatibility condition to ensure that the various charts fit together nicely.

To this end, consider a pair of charts for M, say (U,ϕ) and (V,ψ). Provided these charts overlap, that is, provided $U \cap V \neq \square$, we can form the composite map $\phi\circ\psi^{-1}$. It will map the set

$$\psi(U \cap V) \subseteq R^n \text{ onto the set } \phi(U \cap V) \subseteq R^n$$

as shown in Figure 2.1.2. The map $\phi\circ\psi^{-1}$ is called a *transition* map or an *overlap* map since

$$\phi = (\phi\circ\psi^{-1})\circ\psi \text{ on } U \cap V.$$

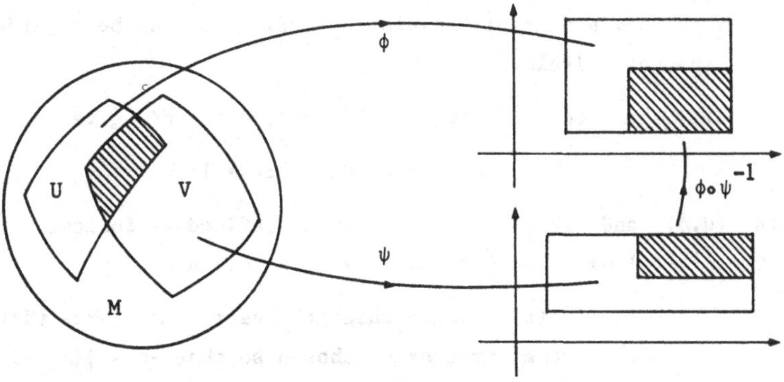

Figure 2.1.2. Overlap Maps

Thus, compatibility as defined below means that we can change charts smoothly. Note also that the definition involves the concept of differentiability in a context where it is already familiar to you — namely, for functions between open sets in R^n.

2.1.2. Definition. By saying that the charts (U, ϕ) and (V, ψ) are *compatible* we mean that if $U \cap V \neq \Box$ then

(i) the sets $\phi(U \cap V)$ and $\psi(U \cap V)$ are open subsets of R^n.

(ii) the map $\phi \circ \psi^{-1}$ is a diffeomorphism between these two sets.

The charts will be called $\mathbb{C}^r$ *compatible* if the overlap map is a diffeomorphism of class $\mathbb{C}^r$ $(1 \leqslant r \leqslant \infty)$. ■

Finally, the long awaited definition of an "atlas", which will contain information about the whole set M:

2.1.3. Definition. An *atlas* for the set M is a collection of mutually compatible charts

$$\{(U_i, \phi_i) \mid i \in I\}$$

such that

$$\bigcup_{i \in I} U_i = M. \quad ■$$

2.1.4. Example. To get some feeling for what has been going on here think of your old school atlas. Extract from it a pair of charts (U, ϕ) and (V, ψ) and then think about the map $\phi \circ \psi^{-1}$. ■

The key to resolving many problems in applied mathematics lies in a suitable choice of "coordinate system". From the modern viewpoint, such co-ordinate systems are simply examples of atlases. The following example shows how a pair of "angular co-ordinates" can be regarded as an atlas on the unit circle.

2.1.5. Example. Let S^1 be the unit circle in R^2 given by

$$S^1 = \{x \in R^2 : \|x\| = 1\}.$$

Charts (U, ϕ) and (V, ψ) for S^1 may be defined as follows:
Put $U = S^1 \setminus \{(-1,0)\}$ and let $\phi: U \rightarrow R$ be given by

$\phi(x)$ = angle which the vector x makes with the first axis, chosen so that $-\pi < \phi(x) < \pi$.

Put $V = S^1\backslash\{(1,0)\}$ and let $\psi : V \to R$ be given by

$\psi(x)$ = angle which the vector x makes with the first axis, chosen so that $0 < \psi(x) < 2\pi$.

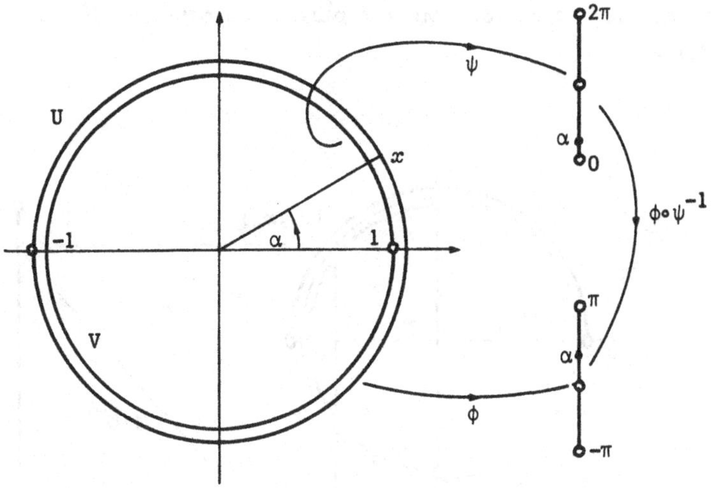

Figure 2.1.3. An atlas for S^1

With the aid of Figure 2.1.3 it may be seen that

$$\phi\circ\psi^{-1} = \begin{cases} \text{id} & \text{on } (0,\pi) \\ \text{id}-2\pi & \text{on } (\pi,2\pi). \end{cases}$$

This map $\phi\circ\psi^{-1}$ is thus a diffeomorphism between the sets

$$\psi(U \cap V) = (0,2\pi)\backslash\{\pi\} \quad \text{and} \quad \phi(U \cap V) = (-\pi,\pi)\backslash\{0\}.$$

Thus the charts (U,ϕ) and (V,ψ) are compatible, where S^1 is covered by U and V. Hence these two charts form an atlas for S^1. ■

EXERCISES 2.1

1. *For each of the following sets, give an atlas containing just one chart:*
   (a) M *is an open set in* R^n
   (b) M *is the graph of a not necessarily continuous function* $f : I \to R$ *where* I *is an open interval on the real line,*
   (c) M *is an* n*-dimensional vector subspace of* R^k $(n \leqslant k)$.

2. *Let* U *consist of all points* (x_1, x_2) *on the unit circle with* $x_2 > 0$ *and let* V *consist of the points with* $x_1 > 0$. *Choose maps*

$$\phi = id_1 |U \quad and \quad \psi = id_2 |V$$

where id_1 *and* id_2 *are the projection maps on* R^2, *as in Figure 2.1.4.*

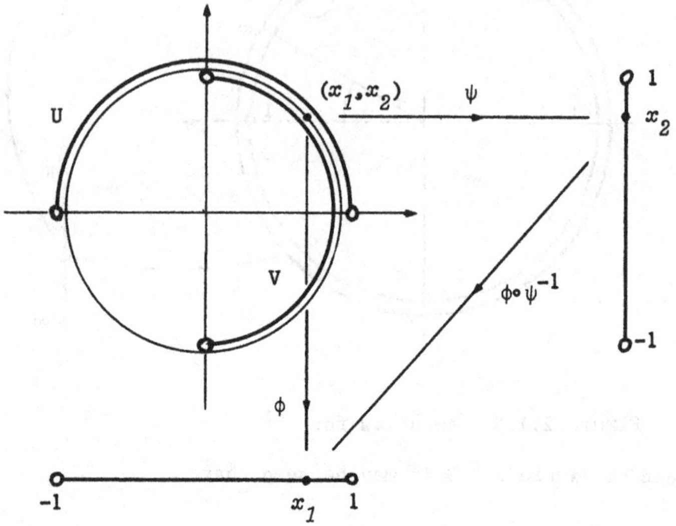

Figure 2.1.4.

(a) *State the domain of* $\phi \circ \psi^{-1}$ *and derive a formula giving its values.*

(b) *Deduce that the charts* (U, ϕ) *and* (V, ψ) *for* S^1 *are compatible.*

(c) *How would you get an atlas containing these two charts?*

3. *Let the sets* U *and* V *be as in Exercise 2 and let maps*

$$\theta : V \to R \quad and \quad \phi : U \to R$$

be defined by putting, for each $x = (x_1, x_2)$ *in the relevant domain,*

$\theta(x)$ = *the angle* α *which the vector* x
makes with the first axis $(-\frac{\pi}{2} < \alpha < \frac{\pi}{2})$,

as indicated in Figure 2.1.5. Let ϕ *be as in question 2 above.*

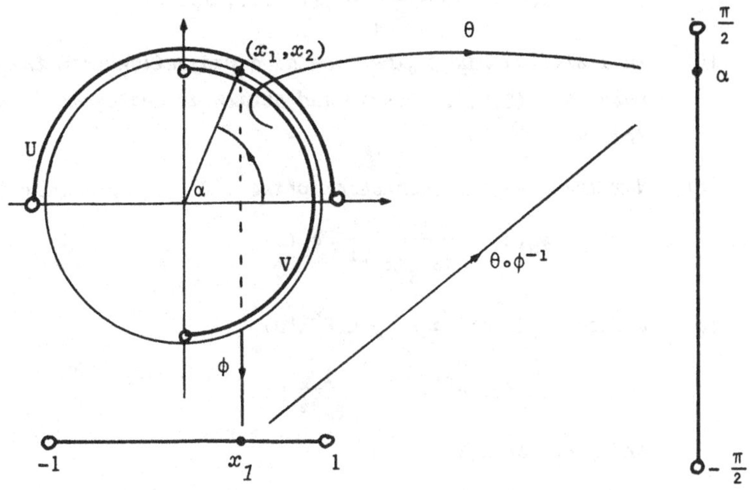

Figure 2.1.5.

(a) *State the domain of* $\theta\circ\phi^{-1}$ *and find a formula for this map.*

(b) *Hence show that the charts* (V,θ) *and* (U,ϕ) *for* S^1 *are compatible.*

4. *The unit sphere* $S^n \subseteq R^{n+1}$.

In this exercise you will construct an atlas, consisting of two charts, for the n*-dimensional unit sphere*

$$S^n = \{(x_1, \ldots, x_{n+1}) \mid x_1{}^2 + \ldots + x_{n+1}{}^2 = 1\} .$$

(a) *Let* ϕ_N *be the map obtained by projecting from the north pole* $N = (0, \ldots, 0, 1)$ *onto the equatorial plane*

$$\{(x_1, \ldots\ldots, x_{n+1}) \mid x_{n+1} = 0\}$$

as indicated in Figure 2.1.6.

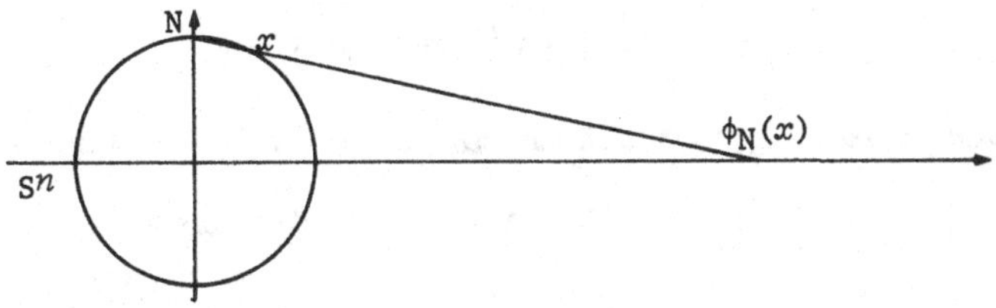

Figure 2.1.6.

Show, by using similar triangles, that for each x *in the set* $S^n\backslash\{N\}$

$$\phi_N(x) = \frac{1}{1 - x_{n+1}} (x_1, \ldots, x_n, 0)$$

(b) *In a similar way define* ϕ_S *as a projection from the south pole* $S = (0, \ldots, 0, -1)$ *and obtain an analogous formula for this map.*

(c) *Now identify the equatorial plane with* R^n *and show that*

$$\phi_N(x) = \frac{1}{\|\phi_S(x)\|^2} \phi_S(x)$$

(d) *Deduce that for each* $y \in R^n \backslash \{0\}$

$$\phi_N \circ \phi_S^{-1}(y) = \frac{1}{\|y\|^2} y$$

and hence show that

$$(S^n \backslash \{N\}, \phi_N) \text{ and } (S^n \backslash \{S\}, \phi_S) .$$

are a pair of charts which form an atlas for S^n.

5. *Show that the map*

$$(r, \theta) : \{(a,b) \in R^2 : a > 0\} \to R^+ \times \left(-\frac{\pi}{2}, \frac{\pi}{2}\right)$$

given by

$$r(a,b) = \sqrt{a^2+b^2}$$
$$\theta(a,b) = \arctan\left(\frac{b}{a}\right)$$

defines a chart for R^2 *which is compatible with the identity chart*

$$\mathrm{id} : R^2 \to R^2 : (a,b) \to (a,b).$$

6. *Let* U *be the subset of the unit sphere* $S^2 \subseteq R^3$ *consisting of the points* (x_1, x_2, x_3) *with* $x_1 > 0$ *and let* V *consist of the points with* $x_3 > 0$. *Let the maps*

$$(\theta, \phi) : U \to R^2 \text{ and } \psi : V \to R^2$$

be given by putting, for all x *in* U *or* V *respectively,*

$$\phi(x) = \alpha \qquad (-\pi < \alpha < \pi)$$
$$\theta(x) = \beta \qquad (0 < \beta < \tfrac{1}{2}\pi)$$
$$\psi(x) = (x_1, x_2)$$

where α *and* β *are the angles shown in Figure 2.1.7.*

(a) *Express* α *and* β *as functions of* $(x_1, x_2, x_3) \in U \cap V$ *and hence find a formula for the map*

$$(\theta,\phi)\circ\psi^{-1} : U \cap V \to R^2 .$$

(b) *Hence show that the charts* $(U,(\theta,\phi))$ *and* (V,ψ) *for* S^2 *are compatible.*

(*Note: the chart* (θ,ϕ) *has been obtained by restricting "spherical polar coordinates "* (r, θ, ϕ) *to part of the unit sphere.*)

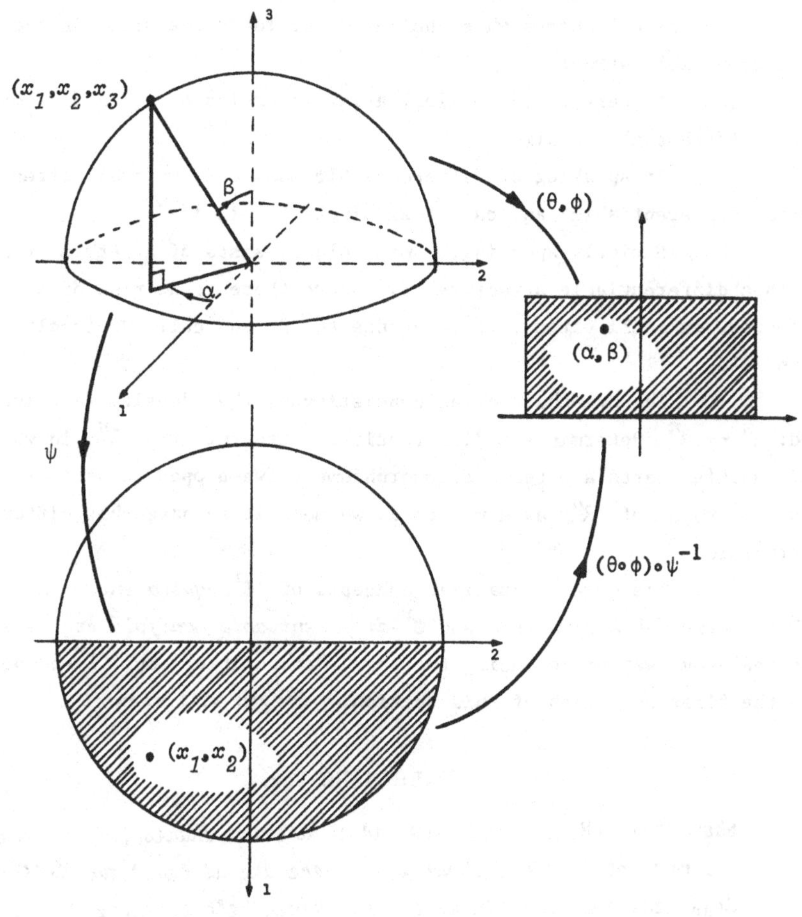

Figure 2.1.7.

2.2. DEFINITION OF A DIFFERENTIABLE MANIFOLD

A given set M may have many different atlases. By saying that an atlas $\mathcal{A}$ for the set M is *equivalent* to another atlas $\mathcal{B}$ for the same set, we mean that every chart in $\mathcal{A}$ is compatible with every chart in $\mathcal{B}$.

It is easy to verify that this defines an equivalence relation on the set of all atlases for the set M. Each equivalence class $\mathcal{S}$ of atlases is said to be a *differentiable structure* for the set M.

A *differentiable manifold* is an ordered pair (M, $\mathcal{S}$) where M is a set and $\mathcal{S}$ is a differentiable structure for M.

The *maximal atlas* of a differentiable manifold (M, $\mathcal{S}$) is the collection of all charts which belong to at least one atlas in the differentiable structure $\mathcal{S}$.

A chart in the maximal atlas is called an *admissible chart* for the differentiable manifold.

In speaking of differentiable manifolds we shall often omit the word "differentiable" and call them simply "manifolds".

Strictly speaking, a manifold consists of a set M together with a differentiable structure $\mathcal{S}$. Where there is no risk of ambiguity, however, we shall often omit reference to $\mathcal{S}$ and call M itself the manifold.

The one-chart atlas consisting of the identity function id: $R^n \to R^n$ determines a differentiable structure for R^n, in which the admissible charts are the diffeomorphisms between open subsets of R^n. When we speak of R^n as a manifold, we mean it to have this differentiable structure.

The more generalized concepts of $\mathcal{C}^r$*-equivalent*, $\mathcal{C}^r$*-differentiable structure* and $\mathcal{C}^r$*-differentiable manifold* are defined in an analogous way on replacing the word "compatible" by "$\mathcal{C}^r$-compatible" in the first paragraph of this section. *Smooth* will mean $\mathcal{C}^\infty$.

EXERCISES 2.2.

1. *Show that* (R, id) *and* (R, id^3) *are two charts for* R *which are incompatible. (This shows that there are at least two different differentiable structures on* R. *Nevertheless, these two structures will later be shown to be "diffeomorphic".)*

2. *Show that the compatibility between charts for a set* M *is not an equivalence relation.*

Check that, nevertheless, equivalence of atlases is an equivalence relation.

3. *Let* $\mathbf{A}$ *be a maximal atlas for a set* M *and let* (U,ϕ) *be a chart in* $\mathbf{A}$.
Consider a set $V \subseteq U$. *Show that* $(V, \phi|V)$ *is also a chart in* $\mathbf{A}$ *if and only if* $\phi(V)$ *is open in* R^n.
(This shows that a maximal atlas must contain a lot of charts).

4. *Dimension of a manifold.*
Let $(M, \mathcal{S})$ *be a differentiable manifold and let* $\mathbf{A}$ *and* $\mathbf{B}$ *be two equivalent atlases in the differentiable structure* $\mathcal{S}$. *Show that if all the charts in* $\mathbf{A}$ *map into* R^n *and all the charts in* $\mathbf{B}$ *map into* R^m *then* $m = n$.
(This integer is called the dimension *of the manifold.)*
(Hint: Let $(U,\phi) \in \mathbf{A}$, $(V,\psi) \in \mathbf{B}$ *with* $U \cap V \neq \square$.
Then consider $D(\phi\circ\psi^{-1})(a)$ *for some* $a \in \psi\ (U \cap V)$.)

5. *Not all manifolds have dimension.*
Show that the subset of R^2 *sketched in Figure 2.2.1. is a manifold by constructing a suitable atlas.*

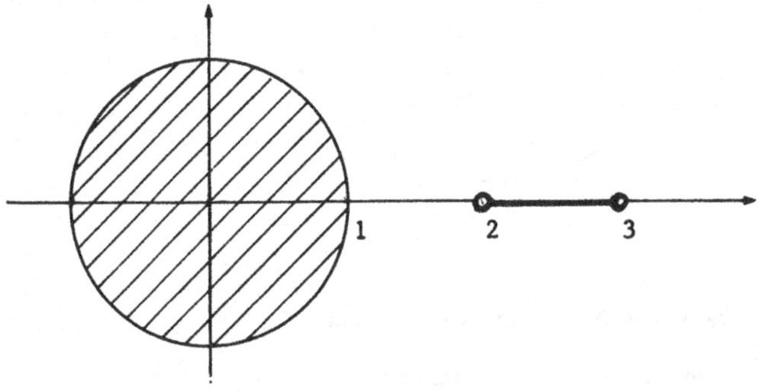

Figure 2.2.1. A disconnected manifold.

6. *Product manifold.*
Let $(M, \mathcal{S})$ *and* $(N, \mathcal{T})$ *be two manifolds.*
Check that the following collection is an atlas for the set $M \times N$:

$$\{(U\times V,\ \phi\times\psi) : (U,\phi) \text{ is an admissible chart for } (M,\mathcal{S}) \text{ and } (V,\psi) \text{ is an admissible chart for } (N,\mathcal{T})\}$$

(The resulting manifold is denoted by $(M\times N,\ \mathcal{S}\times\mathcal{T})$ *and is called the* product *of the two original manifolds.)*

2.3. TOPOLOGIES

We now show how a differentiable manifold can be made into a topological space. This will make it meaningful to speak about the continuity of maps between manifolds.

Before reading further into this section, however, you may wish to revise your topology. See for example Simmons(1963).

2.3.1. Definition. Let M be a differentiable manifold. By saying that a set $A \subseteq M$ is *open* we mean that for each $x \in A$ there is an admissible chart (U,ϕ) for M such that $x \in U$ and $U \subseteq A$. ■

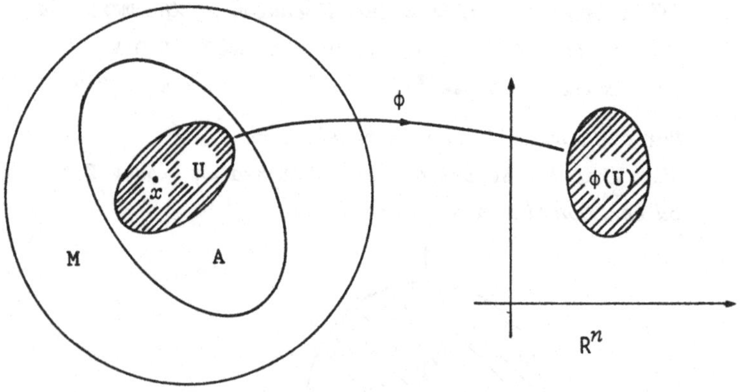

Figure 2.3.1. An open subset of M.

It is easy to verify that the collection of all "open sets" defined in this way satisfies the axioms of a topology for the set M. We shall say that this topology is *induced* by the differentiable structure.

Note that for each chart (U,ϕ) the set U is open in this topology.

We shall suppose that the induced topology is *Hausdorff* and *second countable*, like the usual topology on R^n.

EXERCISES 2.3.

1. *Check that the collection of "open sets" defined on the manifold* M *does indeed satisfy the axioms for a topology.*

2. *Show that a subset of the manifold* M *is open if and only if it can be written as a union of domains of admissible charts.*

3. *Show that for each admissible chart* (U,ϕ) *of* M

   (a) ϕ^{-1} *is continuous*

   (b) ϕ *is continuous .*

   For (b) use Exercise 2.2.3.

4. *Let* M *be an open subset of* R^n *together with the usual differentiable structure, containing the injection map as a chart. Show that the topology induced on* M *is the relative topology.*

5. *Let* S^n *be the unit sphere in* R^{n+1} *with the usual differentiable structure, induced by the polar projections of Exercise 2.1.4. Show that the topology induced on* S^n *is the relative topology. (Hint: Show that the polar projections* ϕ_N *and* ϕ_S *can be extended to continuous maps defined on open subsets of* R^{n+1}*).*

6. *Show that all connected manifolds have dimension.*

3. SUBMANIFOLDS

It is possible to give some quite nasty looking subsets of R^n *the structure of a differentiable manifold as in Exercise 2.1.1(b). In this chapter, however, we shall single out certain "nice" subsets of* R^n *which correspond to our intuitive ideas of smooth curves, surfaces, etc. We shall introduce the idea of a submanifold to describe these goemetric objects precisely.*

A submanifold of R^n *is essentially a subset of* R^n *which can, at least locally, be "flattened out" into a subspace of* R^n. *A useful criterion for a set to be a submanifold of* R^n *is given by the implicit function theorem.*

3.1. WHAT IS A SUBMANIFOLD?

We would like to be able to call a sphere S a "submanifold" of R^3. In Figure 3.1.1. the diffeomorphism ϕ sends the open set $U \subseteq R^3$ into the open set $\phi(U) \subseteq R^3$. At the same time it flattens out the piece of the sphere $S \cap U$ into a subset of the plane $R^2 \times \{0\}$ (which is of course a vector subspace of R^3).

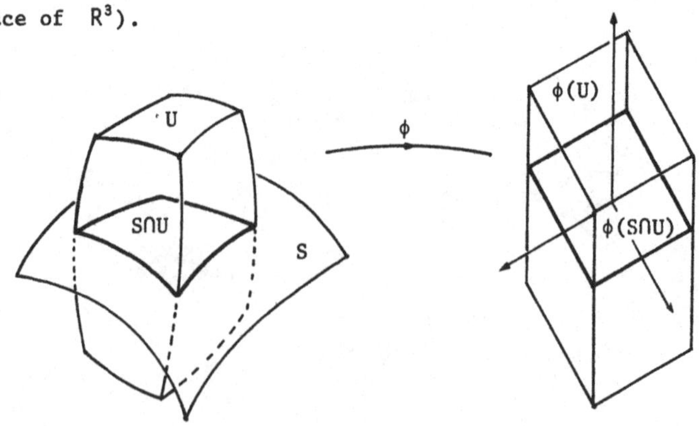

Figure 3.1.1. A submanifold chart for S.

3.1.1. Definition. Let (U, ϕ) be a chart for a manifold M where $\phi : U \to R^n$. By saying that this chart *has the submanifold property for a subset* S *of* M we mean that, for some integer $k \leqslant n$,

$$(\forall\, x \in S \cap U)\ \phi(x) = (\phi_1(x), \ldots, \phi_k(x), 0, \ldots, 0)$$

and, furthermore,

$$\phi(S \cap U) = \phi(U) \cap R^k \times \{0\}. \ ■$$

We would like to be able to claim that the restricted map $\phi|S \cap U$ was a chart for the set S. There is, however, no hope of this if $k < n$ since then the map would not be onto an open set in R^n.

We can easily get over this by dropping the zeros from the end of the n-tuple $\phi(x)$. The image of this restricted map is therefore the set $\phi(S \cap U)$. Since $\phi(U)$ is open in R^n this image set may be identified, on dropping the zeros, with an open set in R^k. In this way, we may regard the map $\phi|S \cap U$ as a chart for the set S.

It may happen that there are enough charts for S of the above form to make up an atlas for S.

3.1.2. Definition. Suppose that S is a subset of a manifold M which has an atlas consisting of charts of the form

$$(S \cap U,\ \Pi \circ \phi|S \cap U)$$

where (U, ϕ) is an admissible chart for M with the submanifold property for S and where $\Pi : R^n \to R^k$ is the natural projection, given by

$$\Pi(x_1, \ldots, x_k, \ldots x_n) = (x_1, \ldots, x_k).$$

The set S, together with the differentiable structure containing these charts, is then called a *submanifold* of M. ■

Note that, as a manifold, S then has dimension k.

EXERCISES 3.1.

1. *Show that each open subset of a manifold* M *is a submanifold of* M.

2. *Let* S *be the set* $\{(x,y) : y = x\}$ *in* R^2 *and let* $\phi : R^2 \to R^2$ *with*

$$\phi(x,y) = (x,\ y-x)$$

Find the image under ϕ *of the set* S.
Show that ϕ *is a diffeomorphism, hence an admissible chart for* R^2.
Deduce that S *is a submanifold of* R^2.

3. *Let* $f: I \to R$ *where* I *is an open interval in* R.
Show that the graph of f *is a submanifold of* R^2 *provided* f *is differentiable.*
(Hint: you need a chart (U, ϕ) *for* R^2 *with the submanifold property for the graph. A picture plus a little experimentation should lead you to a suitable formula for* $\phi(x,y)$ *in terms of* $f(x)$ *and* y.*)*

4. *Prove that every vector subspace of* R^n *of dimension* k *is a submanifold of* R^n *of dimension* k.

5. *Find a differentiable function* $\theta : R^2 \to R^2$ *which maps the first axis onto the graph of the absolute value function* abs *as in Figure 3.1.2.*
(Hint: the next exercise says this can't happen if θ *is a diffeomorphism.)*

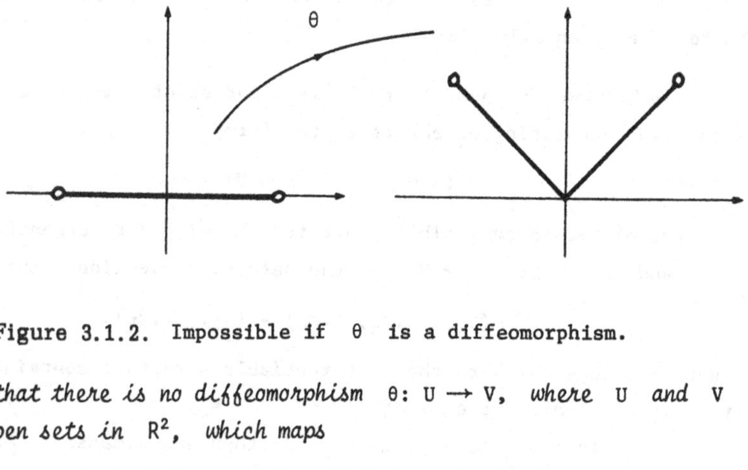

Figure 3.1.2. Impossible if θ is a diffeomorphism.

6. *Show that there is no diffeomorphism* $\theta: U \to V$, *where* U *and* V *are open sets in* R^2, *which maps*

U ∩ first axis *onto* V ∩ graph of abs

Deduce that the graph of abs *cannot be a submanifold of* R^2.

7. *Compare the above exercise with Exercise 2.1.1(b).*
Give an example of the subset of R^2 *which can be given a manifold structure but which nevertheless is not a submanifold of* R^2.

8. *Let* M *be a manifold of dimension* n *and let* S *be a submanifold of* M *of dimension* k. *Suppose that* (U, ϕ) *and* (V, ψ) *are admissible charts for* M *with the submanifold property for* S *and let* $\Pi: R^n \to R^k$ *be the natural projection.*

(a) *Let* $I : \mathbb{R}^k \to \mathbb{R}^n$ *be the natural injection given by*

$$I(x_1,\dots,x_k) = (x_1,\dots,x_k, 0,\dots,0)$$

By drawing a suitable mapping diagram convince yourself that the following equality between maps holds on the domain $\Pi\circ\psi\ (U \cap V \cap S)$:

$$(\Pi\circ\phi)\circ(\Pi\circ\psi)^{-1} = \Pi\circ\phi\circ\psi^{-1}\circ I$$

(b) *Show, via the chain rule, the compatibility of the charts for* S:

$$(U \cap S,\ \Pi\circ\phi|U \cap S) \text{ and } (V \cap S,\ \Pi\circ\psi|V \cap S)\ .$$

(Thus the differentiable structure on a submanifold is unique.)

9. *Show that "being a submanifold of" is a transitive relation.*

10. *Let* S *be a submanifold of* M. *Show that the topology induced on* S *as a submanifold of* M *is the same as the relative topology on* S *as a subset of* M.

3.2. THE IMPLICIT FUNCTION THEOREM

To motivate this theorem, consider the following question: given a function $f: \mathbb{R}^2 \to \mathbb{R}$, *does the set of points*

$$\{(x,y) \mid f(x,y) = 0\}$$

form the graph of some function?

Unfortunately the answer is *no*. Consider the function $f : \mathbb{R}^2 \to \mathbb{R}$ defined by

$$f(x,y) = x^2 + y^2 - 1.$$

The set of points $f^{-1}(0)$ is then just the unit circle in $\mathbb{R}^2$ and this cannot be the graph of any function.

So let's replace the above *global* question by a less ambitious *local* question: given a point $(a,b) \in f^{-1}(0)$, *is there a neighbourhood* $N = I \times J$ *of* (a,b) *such that the set*

$$N \cap f^{-1}(0)$$

forms the graph of some function?

In the case of the circle the answer if *yes* provided the point (a,b) is neither $(1,0)$ nor $(-1,0)$ as is obvious from Figure 3.2.1. Note that at these excluded points the partial derivative $\partial_2 f$ is zero.

The implicit function theorem answers our local question for an arbitrary smooth function $f : \mathbb{R}^{n+m} \to \mathbb{R}^m$.

3.2.1. Theorem. (Implicit function theorem).

Let $f: U \subseteq \mathbb{R}^n \times \mathbb{R}^m \to \mathbb{R}^m$ *where* U *is open and suppose that* f *is* $\mathbb{C}^r$ $(r \geq 1)$.

Suppose there is a point (a,b) *in* U *such that* $f(a,b) = 0$ *and suppose that the linear map*

$$\partial_2 f(a,b) : \mathbb{R}^m \to \mathbb{R}^m$$

is a vector space isomorphism.

There is then a neighbourhood $N = I \times J \subseteq U$ *of* (a,b), *where* I *and* J *are open sets in* $\mathbb{R}^n$ *and* $\mathbb{R}^m$ *respectively, and a function* $g : I \to J$ *such that*

$$\forall\ (x,y) \in I \times J \qquad f(x,y) = 0 \iff y = g(x).$$

The function g *is* $\mathbb{C}^r$.

Proof. See Spivak(1965) or Dieudonné(1960). ■

The following corollary is an important step towards our goal of establishing a criterion for $f^{-1}(0)$ to be a submanifold of $\mathbb{R}^{n+m}$.

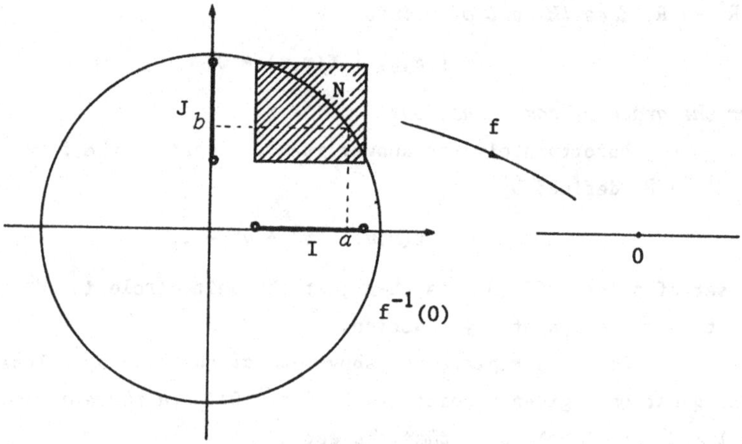

Figure 3.2.1.

3.2.2. Corollary. *Let the map* f *and the neighbourhood* N *of* (a,b) *be as in Theorem 3.2.1. There is then a map* $\phi : N \to \mathbb{R}^{n+m}$ *such that* (N,ϕ) *is a chart for* $\mathbb{R}^{n+m}$ *with the submanifold property for* $f^{-1}(0)$.

Proof. This homes in on the fact that the set $f^{-1}(0) \cap N$ is the graph of a function. Details are left as an exercise. ■

EXERCISES 3.2.

1. *Use the implicit function theorem to prove the (geometrically obvious) claims made at the beginning of section 3.2. for the example*

$$f(x,y) = x^2 + y^2 - 1.$$

2. *Let* $g : I \to J$ *be* C^r *where* I *and* J *are open sets in* R^n *and* R^m *respectively. With a little experimentation you should be able to guess a formula for a map* $\phi : I \times J \to R^n \times R^m$ *which*

   (a) *maps the graph of* g *onto* $I \times \{0\}$ *as in Figure 3.2.2.*

   (b) *is a diffeomorphism from* $I \times J$ *onto an open set in* $R^n \times R^m$.

   Hence prove Corollary 3.2.2.

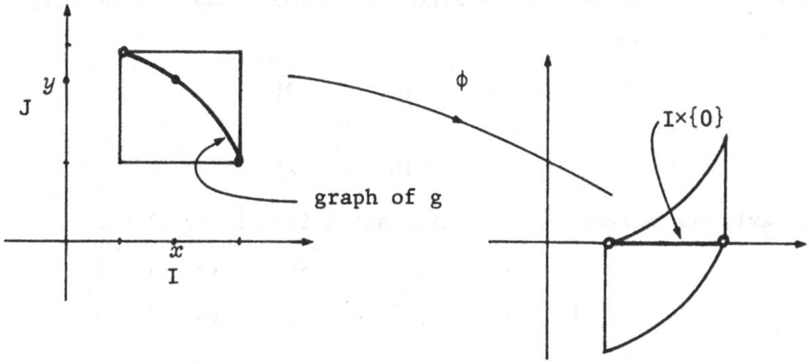

Figure 3.2.2.

3.3. A TEST OF SUBMANIFOLDS

The following theorem provides a useful criterion for showing that certain sets are submanifolds of R^{n+m}.

3.3.1. Theorem. *Let* $f : U \subset R^n \times R^m \to R^m$ *where* U *is open and suppose that* f *is* C^r $(r \geq 1)$.

If at every point (a,b) *of* $f^{-1}(0)$ *the linear map*

$$D\,f(a,b) : R^{n+m} \to R^m$$

has rank m, *then the set* $f^{-1}(0)$ *is a submanifold of* R^{n+m} *whose dimension is* n.

The idea of the proof is to use Corollary 3.2.2. of the implicit function theorem to get a chart defined on a neighbourhood of the point (a,b) with the submanifold property for $f^{-1}(0)$. Before applying this corollary, however, we may need to "change co-ordinates" to ensure that the point (a,b) satisfies the hypothesis relating to $\partial_2 f(a,b)$.

The desired change of co-ordinates will be constructed with the aid of a technical result from matrix theory. This involves the use of a permutation matrix to move the columns of a matrix. Consider the following simple example:

$$\begin{pmatrix} 1 & 2 & 4 \\ 3 & 0 & 0 \end{pmatrix}$$

This matrix has rank 2, that is 2 of its columns are linearly independent (for example, $\begin{pmatrix} 4 \\ 0 \end{pmatrix}$ and $\begin{pmatrix} 1 \\ 3 \end{pmatrix}$). If we want to "shift all of the rank" into the last two columns via a linear diffeomorphism, we could apply the map whose matrix is

$$\begin{pmatrix} 0 & 1 & 0 \\ 1 & 0 & 0 \\ 0 & 0 & 1 \end{pmatrix}$$

Clearly the determinant of this matrix is -1 $(\neq 0)$ and

$$\begin{pmatrix} 1 & 2 & 4 \\ 3 & 0 & 0 \end{pmatrix} \begin{pmatrix} 0 & 1 & 0 \\ 1 & 0 & 0 \\ 0 & 0 & 1 \end{pmatrix} = \begin{pmatrix} 2 & 1 & 4 \\ 0 & 3 & 0 \end{pmatrix}$$

as required. What we have achieved here is a shifting of all of the rank into the right hand columns by using a 3×3 matrix whose columns consist of the linearly independent vectors e_2, e_1 and e_3.

The 2×2 submatrix

$$\begin{pmatrix} 1 & 4 \\ 3 & 0 \end{pmatrix}$$

obtained from picking out those last two columns is the matrix of a vector space isomorphism from R^2 to R^2.

Now as an exercise see if you can construct a 4×4 matrix similar to the one above which "moves all the rank" of

$$\begin{pmatrix} 3 & 2 & 1 & 1 \\ 4 & 4 & 2 & 1 \\ 1 & 2 & 1 & 1 \end{pmatrix}$$

to the right (the answer is not unique).

What is required is the following lemma:

3.3.2. Lemma. *Let* A *be an* $m\times(n+m)$ *matrix of rank* m. *Then there exists an* $(n+m)\times(n+m)$ *matrix* Q, *all of whose columns are linearly independent and consist of the vectors* $e_1, e_2, \ldots, e_{n+m}$ *(not necessarily in that order) with the property that*

$$A\,Q\,P$$

is a nonsingular $m\times m$ *matrix and* P *is the* $(n+m)\times m$

$$\begin{bmatrix} 0 & 0 & \cdots & 0 \\ \vdots & \vdots & & \vdots \\ 0 & 0 & \cdots & 0 \\ \hline & I_{m\times m} & & \end{bmatrix}$$

(Here P *has the effect of picking out the required* $m\times m$ *submatrix).*

The proof is just an exercise in elementary matrix theory. ■

EXERCISES 3.3.

1. *Use Theorem 3.3.1. to prove that the unit sphere* S^n *is a submanifold of* R^{n+1} *of dimension* n.

2. *Let* $O(n)$ *be the set of all* $n\times n$ *orthogonal matrices regarded as a subset of* $R^{n\times n}$. *Prove that* $O(n)$ *is a submanifold of* $R^{n\times n}$ *of dimension* $\frac{1}{2}n(n-1)$.

3. *Let* $SO(n)$ *be the set of all* $n\times n$ *rotation matrices, that is, orthogonal matrices with determinant* 1. *Show that* $SO(n)$ *is an open subset of* $O(n)$. *Hence show that* $SO(n)$ *is also a submanifold of* $R^{n\times n}$ *of dimension* $\frac{1}{2}n(n-1)$.

4. *Here we prove Theorem 3.3.1. by using Lemma 3.3.2.*

   (a) *Use the lemma to show that there is a diffeomorphism*

   $$q: R^{n+m} \to R^{n+m} \quad \text{such that} \quad \partial_2 f\circ q(q^{-1}(a,b))$$

   is a vector space isomorphism from R^m *to* R^m.

   (b) *Now apply Corollary 3.2.2. of the implicit function theorem to the map* $f\circ q$ *to deduce the existence of a chart which is defined on a neighbourhood of the point* $q^{-1}(a,b)$ *and has the submanifold property for* $(f\circ q)^{-1}(0)$.

(c) *With the aid of Figure 3.3.1. deduce the existence of a chart which is defined on a neighbourhood of the point* (a,b) *and has the submanifold property for* $f^{-1}(0)$.

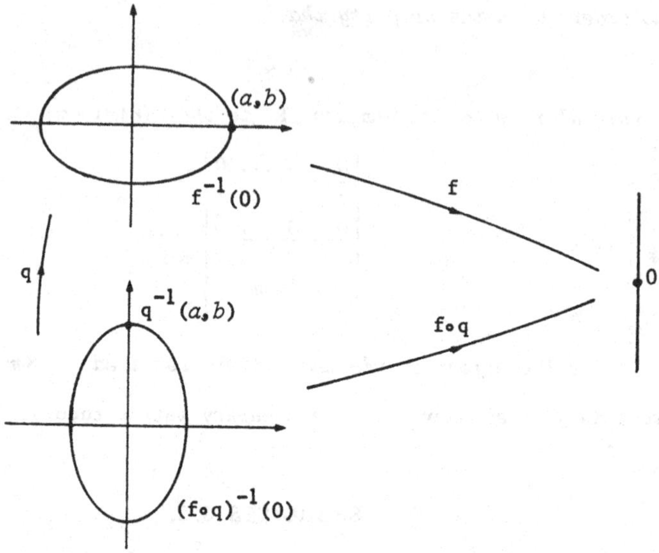

Figure 3.3.1.

3.4. ROTATIONS IN R^3

Exercise 3.3.3. states that $SO(n)$ is a submanifold of $R^{n\times n}$ of dimension $\frac{1}{2}n(n-1)$. The set of 3×3 rotation matrices $SO(3)$ is important in the study of rigid body motion, which will be dealt with in detail in Chapter 14. A summary of the results which will be needed is given below.

3.4.1. **Lemma.** $SO(3)$ *is a 3-dimensional submanifold of* $R^{3\times3}$.

Proof. This is a part of Exercise 3.3.3. ∎

3.4.2. **Lemma.** $SO(3)$ *under the operation of matrix multiplication is a group.*

Proof. Elementary matrix theory shows that $SO(3)$ is closed under matrix multiplication and matrix inversion and that it contains the identity. ∎

We will denote the linear map with matrix A relative to the usual bases for R^3 by L_A. Thus for each $a \in R^3$

$$L_A(a) = Aa$$

where on the right we regard a as a column matrix. We will call the map L_A a *rotation* for reasons which will soon be apparent.

3.4.3. Lemma. *Rotations preserve distance.*

Proof. Let $a \in R^3$ then for $A \in SO(3)$

$$\begin{aligned} \|a\|^2 &= a^t a \\ &= a^t A^t A a \quad \text{since } A^t A = I \\ &= (Aa)^t (Aa) \\ &= \|Aa\|^2 . \end{aligned}$$

Thus for $a,b \in R^3$

$$\begin{aligned} \|a-b\| &= \|A(a-b)\| \\ &= \|Aa - Ab\| \quad \text{as required. } \blacksquare \end{aligned}$$

3.4.4. Definition. The *orientation* of the ordered triple of vectors (a_1,a_2,a_3), (b_1,b_2,b_3) and (c_1,c_2,c_3) based at the same point in R^3 is given by the sign of the determinant

$$\det \begin{bmatrix} a_1 & b_1 & c_1 \\ a_2 & b_2 & b_2 \\ a_3 & b_3 & c_3 \end{bmatrix}. \quad \blacksquare$$

Note that the determinant in the above definition is simply the scalar triple product $(a_1,a_2,a_3) \times (b_1,b_2,b_3).(c_1,c_2,c_3)$.

3.4.5. Lemma. *Rotations preserve orientations.*

Proof. Let $A \in SO(3)$. The orientation of any triple of vectors

$$(x_1,x_2,x_3),\ (y_1,y_2,y_3),(z_1,z_2,z_3)$$

in R^3 is the sign of

$$\det \begin{bmatrix} x_1 & y_1 & z_1 \\ x_2 & y_2 & z_2 \\ x_3 & y_3 & z_3 \end{bmatrix} = \det X, \text{ say,}$$

while after rotation their orientation is the sign of

$$\begin{aligned} \det \left[A \begin{pmatrix} x_1 \\ x_2 \\ x_3 \end{pmatrix} \; A \begin{pmatrix} y_1 \\ y_1 \\ y_1 \end{pmatrix} \; A \begin{pmatrix} z_1 \\ z_2 \\ z_3 \end{pmatrix} \right] &= \det AX \\ &= \det A \det X \\ &= \det X, \text{ as } \det A = 1. \end{aligned}$$

Thus the orientation is the same. $\blacksquare$

Another useful property of elements of $SO(3)$ is as follows:

3.4.6. Lemma. *A matrix* $A \in SO(3)$ *is uniquely determined by its effect on a linearly independent pair of vectors in* R^3.

Proof. Let A and A' be two matrices in $SO(3)$ which map a linearly independent pair of vectors a,b in R^3 into the same pair of vectors x,y respectively, which must also be linearly independent. Since rotations preserve distance and orientation, $a \times b$ must be mapped to $x \times y$ in each case. Thus A and A' have the same effect on the basis $a, b, a \times b$ for their domain and hence must be identical. ∎

EXERCISES 3.4.

1. *Let* $A \in SO(3)$ *and let* L_A *be the corresponding linear map.*
   (a) *Prove that* L_A *maps the unit sphere* S^2 *in* R^3 *onto itself.*
   (b) *Prove that the dot product* $a.b = a^Tb$ *stays the same when the the two vectors* a *and* b *are mapped under* L_A.

2. *Let* $A \in SO(3)$. *Prove that if* λ *is a complex eigenvalue of* A *and* x *a corresponding eigenvector while* $\bar{\lambda}$ *and* $\bar{x}$ *are their complex conjugates, then*

$$\bar{x}^T x = \bar{\lambda}\lambda\, \bar{x}^T x$$

hence

$$\bar{\lambda}\lambda = 1.$$

3. *Use the fact that the product of the eigenvalues of a matrix (each counted with appropriate multiplicity) is equal to the determinant of the matrix to prove that every matrix* $A \in SO(3)$ *has an eigenvalue equal to* 1. *(Physically, the eigenspace corresponding to the eigenvalue* 1 *will be the axis of rotation of a rigid body.)*

4. DIFFERENTIABILITY

Differentiable manifolds provide the natural context in which to discuss the differentiability of functions.

In advanced calculus differentiability was studied for maps between open sets in normed vector spaces. In this chapter the concept of differentiability will be extended so as to apply to maps between differentiable manifolds.

4.1. LOCAL REPRESENTATIVES

A manifold is a set which can be locally identified with an open subset of a Euclidean space.

This property enables us to define smoothness properties of functions between arbitrary manifolds.

4.1.1. Definition. Let M and N be two differentiable manifolds and let $f : M \to N$. Suppose furthermore that (U, ϕ) and (V, ψ) are admissible charts for M and N respectively, with $f(U) \subseteq V$. The map

$$\psi \circ f \circ \phi^{-1} : \phi(U) \to \psi(V)$$

is called the *local representative of* f *relative to the two given charts* and will be denoted by $f_{\phi\psi}$. ■

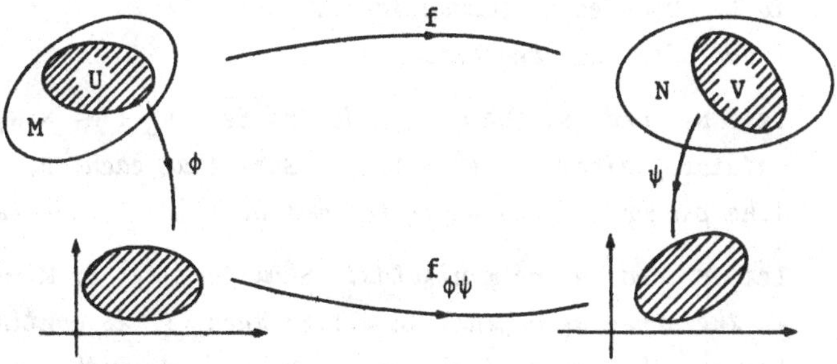

Figure 4.1.1. Local representatives.

Since the local representative $f_{\phi\psi}$ maps an open set in some Euclidean space into another such set it makes sense to speak about its differentiability, as understood in advanced calculus. Hence the following definition.

4.1.2. Definition. Let $f: M \to N$ and let $a \in M$. By saying that f is *differentiable at* a we mean:

For each admissible chart (V, ψ) for N with $f(a) \in V$ there is an admissible chart (U, ϕ) for M with $a \in U$ such that $f(U) \subseteq V$ and the local representative $f_{\phi\psi}$ is differentiable at $\phi(a)$. ■

Note that the definition has been framed in such a way that differentiability of f will ensure its continuity.

By saying that f is *differentiable on a subset* S of M we mean that f is differentiable at every point of S. By saying that f is *differentiable* we mean that it is differentiable on M.

We shall also assume the analogous definitions in which the word "differentiable" is everywhere replaced by "is of class $\mathfrak{C}^r$ $(r \geq 1)$". *Smooth* means $\mathfrak{C}^\infty$.

EXERCISES 4.1.

1. *Let* L, M *and* N *be manifolds. Show that if* $f: L \to M$ *and* $g: M \to N$ *are* $\mathfrak{C}^r$ $(r \geq 1)$ *then so is* $g \circ f : L \to N$.

2. *Let* M *be a set and let* N *be a manifold. Suppose, furthermore, that there is a bijective map* $f: M \to N$. *Show how to define an atlas for the set* M *which makes the map* f *to be* $\mathfrak{C}^r$.

3. *Let* M *and* N *be manifolds. Show that if* $f: M \to N$ *is* $\mathfrak{C}^r$ *then for each open set* $U \subset M$ *and each open set* $V \supset f(U)$ *in* N, *the restricted map* $f|U : U \to V$ *is also* $\mathfrak{C}^r$, *where the open sets are to be regarded as submanifolds.*
Prove also the converse.

4. *Let* N_1 *and* N_2 *be manifolds and let* $\Pi_i : N_1 \times N_2 \to N_i$ *be the natural projection* $(i = 1,2)$. *Show that each* Π_i *is* $\mathfrak{C}^r$.
(The product of two manifolds was defined in exercise 2.2.6.)

5. *Let* M *and* N *be manifolds. Show that if* $f: M \to N$ *is continuous in the sense of Definition 4.1.2, then* f *is continuous as a map between the topological spaces* M *and* N *with topologies given by Definition 2.3.1.*

4.2. MAPS TO OR FROM REAL SPACES

We may regard R^n as a manifold with its usual differentiable structure containing the identity map as a chart. Thus if M is any manifold it makes sense now to speak about the differentiability of a map $f: M \to R^n$ or a map $g: R^n \to M$.

For such maps the definition given in the previous section may be reduced to a somewhat simpler criterion, as follows:

4.2.1. Theorem. *Let* $f: M \to R^n$ *and let* $a \in M$. *The map* f *is* $\mathfrak{C}^r$ *at* a *if and only if for each open set* V *in* R^n *with* $f(a) \in V$ *there is an admissible chart* (U, ϕ) *for* M *with* $a \in U$ *such that*

(i) $f(U) \subseteq V$

(ii) *the map* $f\circ\phi^{-1} : \phi(U) \to R^n$ *is of class* $\mathfrak{C}^r$ *at* $\phi(a)$.

The proof of this result is left as one of the exercises, along with the formulation of an analogous result for the differentiability of a map $g : R^n \to M$. See Figure 4.2.1. ■

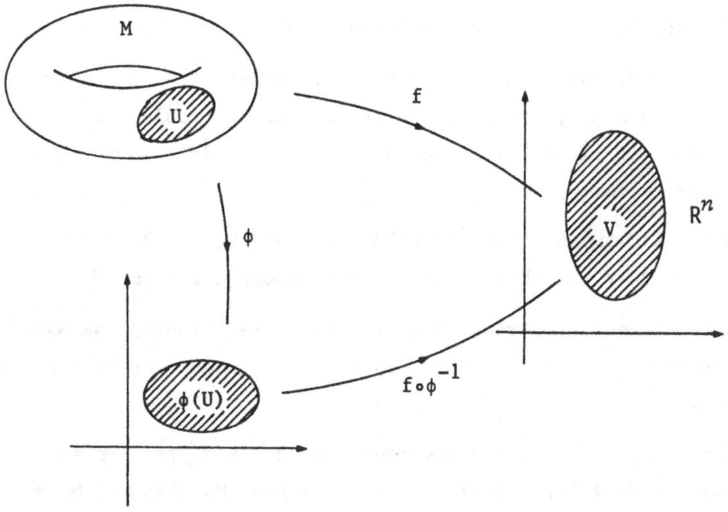

Figure 4.2.1.

EXERCISES 4.2.

1. *Let* X *and* Y *be open sets in* R^m *and* R^n *respectively, so that* X *and* Y *can be regarded as manifolds.*

   Show that a map $f : X \to Y$ *is differentiable as a map between manifolds if and only if it is differentiable as a map between open sets.*

2. *Prove theorem 4.2.1.*

3. *Formulate a theorem analogous to the one in the text for a map* $g : X \to M$ *where* M *is a manifold and* X *is an open set in* R^n. *Draw a diagram and prove your theorem.*

4. (a) *Let* M *be a manifold. Show that a map* $f : M \to R^n$ *is* $\mathbf{C}^r$ *if and only if each of its component functions* $f_i : M \to R$ *is* $\mathbf{C}^r$.

   (b) *Let* S *be a submanifold of* M *and suppose* $f : M \to N$ (N *a manifold*) *is* $\mathbf{C}^r$.

   Show that $f|S : S \to N$ *is also* $\mathbf{C}^r$.

4.3. DIFFEOMORPHISMS

Now that we have a definition of differentiability of functions which map manifolds to manifolds we are in a position to extend the idea of a diffeomorphism to include such functions.

4.3.1. Definition. A map $f: M \to N$, where M and N are manifolds is said to be a *diffeomorphism* if it is differentiable and has a differentiable inverse. If f and its inverse are $\mathbf{C}^r$ it is said to be a $\mathbf{C}^r$ *diffeomorphism.* ■

4.3.2. Definition. Two manifolds M and N are said to be *diffeomorphic to each other* if there exists a diffeomorphism from M to N. ■

We shall regard two manifolds as being "essentiablly the same" if they are diffeomorphic to each other. This idea is summed up by the following lemma.

4.3.3. Lemma. *Suppose the manifold* M *is diffeomorphic to the manifold* N *and* $\{(U_i, \phi_i) \mid i \in I\}$ *is an atlas for* M. *Let* $f: M \to N$ *be a diffeomorphism.*

Then $\{(f(U_i), \phi_i \circ f^{-1}) \mid i \in I\}$ *is an atlas for* N.

Proof. Follows from the definition of a chart. ■

The above and the following lemma will be useful when we deal with the motion of a rigid body.

4.3.4. Lemma. Let $L : R^n \to R^m$ be a linear one-to-one map (where $m \geq n$). Suppose M is a submanifold of R^n.

Then L(M) is a submanifold of R^m which is diffeomorphic to M.

Proof. See exercises. ■

We now mention some celebrated results referred to by Spivak(1970) volume 1, chapter 2 and Stern(1983).

There are, up to diffeomorphism, unique differentiable structures on each R^n if $n \neq 4$. There are at least three "exotic" differentiable structures on R^4 and it is speculated that there are uncountably many.

There are, up to diffeomorphism, unique differentiable structures on S^n for $n \leq 6$. But on S^7 there are 28 essentially different differentiable structures, and for S^{31} there are over 16 million.

EXERCISES 4.3.

1. *Show that if* (U, ϕ) *is an admissible chart for a manifold* M *then* ϕ *is a diffeomorphism from* U *to* $\phi(U) \subset R^n$. *Compare Exercise 2.3.3.*

2. *The two charts* (R, id) *and* (R, id^3) *generate two distinct differentiable structures for* R *as these charts are not compatible. Show that nevertheless the two resulting manifolds are diffeomorphic.*

3. *Prove Lemma 4.3.4.*

5. TANGENT SPACES AND MAPS

In the previous chapter the idea of differentiability was generalized so as to apply to maps between manifolds. In this chapter the idea of the derivative itself will be generalized to this context.

In order to retain the idea of differentiation as a process of linearization it will be convenient to introduce, at each point of a manifold, a vector space known as the tangent space at that point. These vector spaces will supply the domains and codomains for the linearized maps. A neat formulation of the chain rule then becomes possible via a generalization of the tangent functor, from Section 1.2, to manifolds.

Although all the concepts of this chapter can be formulated for arbitrary manifolds, our discussion will be restricted to submanifolds of R^n*. This will provide an adequate background for later chapters on mechanics while avoiding unnecessary abstraction.*

5.1. TANGENT SPACES

In elementary geometry one studies tangents to circles and tangent planes to spheres. The definition of a tangent space will generalize these ideas to arbitrary curves and surfaces and their higher dimensional analogues.

Let M be a submanifold of R^n and let $a \in M$. Intuitively, we want the tangent space at a to consist of all the arrows emanating from a in directions which are "tangential" to M at a, as in Figure 5.1.1. The arrows may be regarded as elements of the vector space T_aR^n, introduced in Section 1.2. To capture the idea of "tangency" we use parametrized curves in the manifold.

5.1.1. Definition. *A parametrized curve* in M at a is a map $\gamma: I \to M$ with $\gamma(0) = a$, where I is an open interval containing 0. The image $\gamma(I)$ of the map γ will be called a *curve* in M at a and γ will be said to *parametrize this curve.*

The parametrized curve will be called *differentiable* (or *of class* C^r) if the map γ is differentiable (or of class C^r). ■

Since $M \subseteq R^n$, the notation in Corollary 1.4.4 is applicable to γ and can be interpreted as follows: If $\gamma(t)$ is the position of a particle moving along the manifold at time t then $\gamma'(t)$ is the velocity of the particle. Intuition requires this to be in a direction "tangential" to the manifold and leads to the following definition, illustrated in Figure 5.1.2.

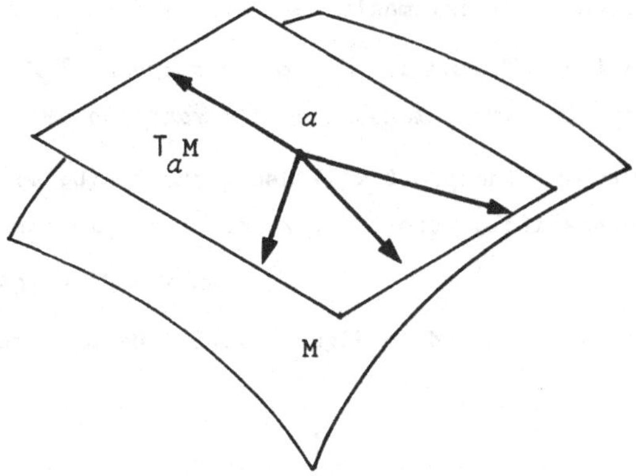

Figure 5.1.1. A tangent space at a.

5.1.2. **Definition.** A *tangent vector* to M at a is an element of T_aR^n of the form $(a, \gamma'(0))$ where γ is a smooth parametrized curve in M based at a. For brevity, we shall sometimes denote this tangent vector by $[\gamma]_a$. ■

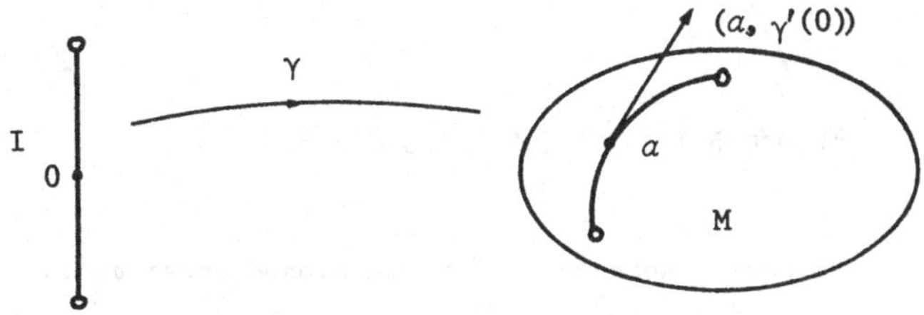

Figure 5.1.2. A tangent vector.

5.1.3. Definition. The *tangent space* to M at a is the set of all tangent vectors to M at a and will be denoted by T_aM. The *tangent bundle* of M is the union of all its tangent spaces and will be denoted by TM so that

$$TM = \bigcup_{a \in M} T_aM. \blacksquare$$

The definition of T_aM implies that it is a subset of the vector space T_aR^n from Section 1.2. The following theorem confirms the expectation that the tangent space should be "flat", rather than "curved" like a typical manifold.

5.1.4. Theorem. *The tangent space* T_aM *is a vector subspace of* T_aR^n *of the same dimension as the manifold* M.

Proof. The proof will use a chart (U, ϕ) for R^n at a which has the submanifold property for M. Hence, in the notation of Section 3.1,

$$\phi(M \cap U) = R^k \times \{0\} \cap \phi(U)$$

as illustrated in Figure 5.1.3. We may assume that $\phi(a) = 0$.

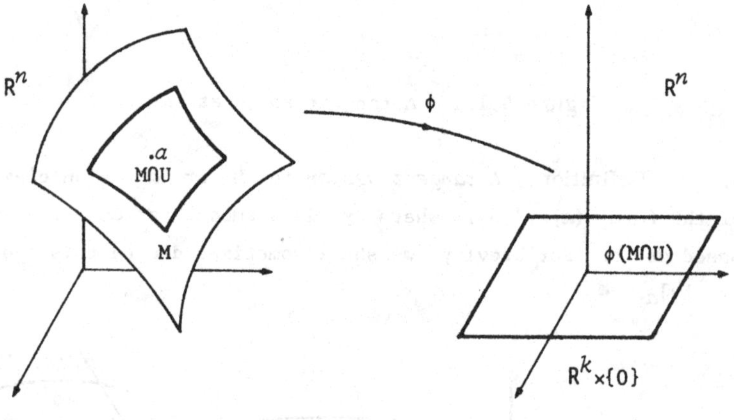

Figure 5.1.3. A submanifold chart.

We shall check that T_aM is closed under vector addition and leave it as an exercise to prove that it is closed under scalar multiplication. So let $(a, \gamma'(0))$ and $(a, \delta'(0))$ be an arbitrary pair of vectors in T_aM, with γ and δ parametrized curves in M based at a. As ϕ is a submanifold chart it follows that both

$$\phi \circ \gamma \quad \text{and} \quad \phi \circ \delta \quad \text{map into} \quad \mathbb{R}^k \times \{0\}$$

and hence also

$$(\phi \circ \gamma)' \quad \text{and} \quad (\phi \circ \delta)' \text{ map into} \quad \mathbb{R}^k \times \{0\}.$$

From Corollary 1.4.4 this implies that

$$D(\phi \circ \gamma)(0)(1) \quad \text{and} \quad D(\phi \circ \delta)(0)(1) \in \mathbb{R}^k \times \{0\}$$

and hence by the chain rule, Theorem 1.1.4,

$$D\phi(a)(\gamma'(0)) \quad \text{and} \quad D\phi(a)(\delta'(0)) \in \mathbb{R}^k \times \{0\}.$$

Hence by linearity of $D\phi(a)$

$$D\phi(a)(\gamma'(0) + \delta'(0)) \in \mathbb{R}^k \times \{0\}.$$

The vectors involved are shown in Figure 5.1.4.

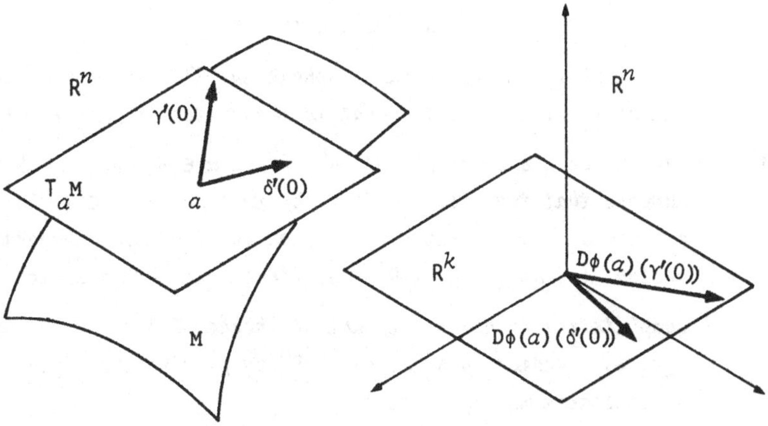

Figure 5.1.4.

We may thus define a map ε from an interval into M by putting

$$\varepsilon(t) = \phi^{-1}(tD\phi(a)(\gamma'(0) + \delta'(0))).$$

Since ε is a curve in M at a, the vector $(a, \varepsilon'(0))$ is an element of T_aM. By the chain rule, however,

$$\begin{aligned} \varepsilon'(0) &= D\phi^{-1}(0) \circ D\phi(a)(\gamma'(0) + \delta'(0)) \\ &= \gamma'(0) + \delta'(0) \end{aligned}$$

and so, by the definition of vector addition in $T_a\mathbb{R}^n$,

$$(a, \gamma'(0)) + (a, \delta'(0)) = (a, \varepsilon'(0)) \in T_a M .$$

Thus $T_a M$ is closed under vector addition.

The proof that $T_a M$ is closed under scalar multiplication is similar and is left as an exercise, as is the statement that $T_a M$ has the same dimension as M. ∎

EXERCISES 5.1.

1. *Let* M *be the unit circle* $\{(x_1,x_2): x_1{}^2 + x_2{}^2 = 1\}$ *in* R^2 *and let* $a = (a_1,a_2) \in M$. *Show that if* $\gamma = (\gamma_1,\gamma_2)$ *is a parametrized curve in* M *based at* a *then*

$$\gamma_1(0)\,\gamma_1{}'(0) + \gamma_2(0)\,\gamma_2{}'(0) = 0 .$$

Deduce that each vector in $T_a M$ *is orthogonal to the vector* (a,a) *in* $T_a R^2$ *and illustrate with a sketch.*

2. *Show that if* M *is the unit sphere in* R^n *and* $a \in M$ *then each vector in* $T_a M$ *is orthogonal to the vector* (a,a) *in* $T_a R^n$.

3. *Let* M *be a submanifold of* R^n, *let* $a \in M$ *and let* $b \in R^n$. *Suppose that the distance from* b *to a point* $x \in M$ *has a minimum at* $x = a$. *Prove that every vector in* $T_a M$ *is orthogonal to the vector* $(a, a-b)$ *in* $T_a R^n$ *and illustrate with a sketch.*

4. *Complete the proof of the part of Theorem 5.1.4 which says that* $T_a M$ *is a vector subspace of* $T_a R^n$ *by proving the closure of* $T_a M$ *under scalar multiplication.*

5. *Let* M *be a submanifold, and* U *an open subset, of* R^n. *Show that* $M \cap U$ *is a submanifold of* R^n *and that* $TM \cap TU = T(M \cap U)$.

6. *Let* M *be a submanifold of* R^n *and let* $a \in M$. *Prove that if* (U,ϕ) *is a submanifold chart for* M *at* a *as in the proof of Theorem 5.1.4, then*

$$T_a M = \{a\} \times D\phi^{-1}(0)(R^k \times \{0\}) .$$

Do this by showing (i) RHS $\subseteq$ *LHS and (ii) LHS* $\subseteq$ *RHS. Techniques used in the proof of Theorem 5.1.4 will come in handy.*

7. *Use the result of Exercise 6 to show that the tangent space* $T_a M$ *has the dimension* k *(which is also the dimension of* M*), thereby completing the proof of Theorem 5.1.4.*

5.2. TANGENT MAPS

The idea of a "local linear approximation" to a map between manifolds can now be formulated in a precise way with the aid of tangent vectors. To this end let $f: M \longrightarrow N$ be differentiable, where M and N are submanifolds of R^m and R^n, and let $a \in M$.

Intuitively, a sufficiently small piece of a curve in a manifold will be indistinguishable from a line segment and hence will correspond to a tangent vector to the manifold. The map f sends a small piece of curve through a in M to another small piece of curve through $f(a)$ in N, as illustrated on the left of Figure 5.2.1.

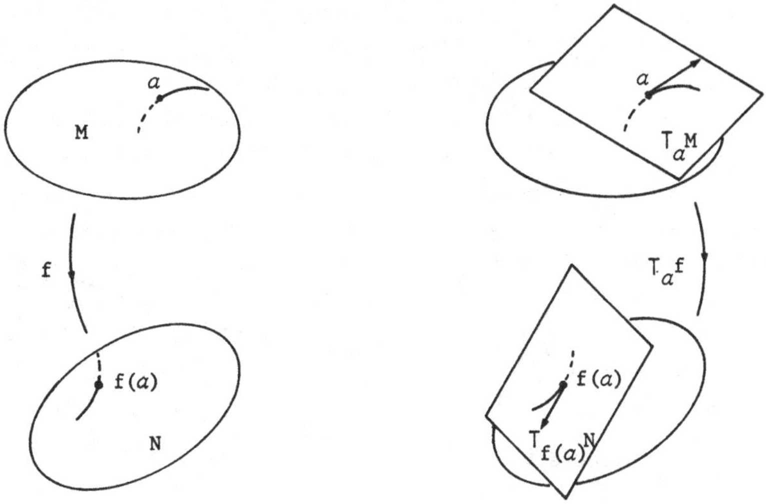

Figure 5.2.1. The tangent map of f at a.

By now imagining the pieces of curve to be so small that we can regard them as tangent vectors, we get the map, shown in the figure as T_af, which sends tangent vectors in T_aM across to tangent vectors in $T_{f(a)}M$. This map is, in a sense, an approximation to f near a and, hopefully, it will be linear.

By introducing parametrizations for these curves as in Figure 5.2.2 and using their derivatives to define the tangent vectors, we are led to propose the following definition.

5.2.1. Definition. The *tangent map of* f *at* a is the map

$$T_af : T_aM \rightarrow T_{f(a)}N$$

with

$$T_a f(a,\gamma'(0)) = (f(a),(f\circ\gamma)'(0))$$

or, equivalently, $$Tf([\gamma]_a) = [f\circ\gamma]_{f(a)}$$

for each differentiable parametrized curve γ in M at a. The *tangent map of* f is the map

$$Tf\colon TM \to TN$$

given by

$$Tf|T_aM = T_af \quad \text{for each } a \in M. \quad \blacksquare$$

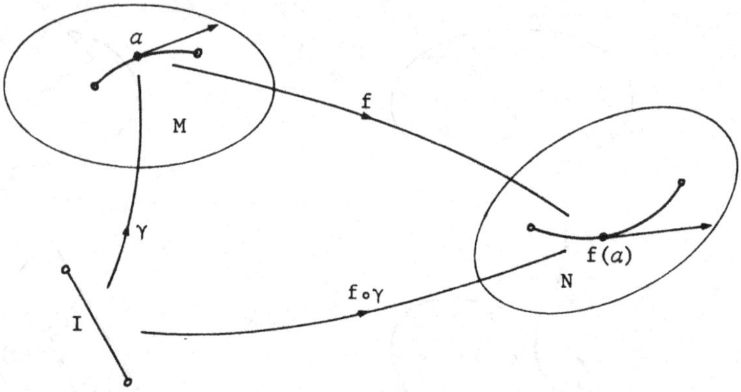

Figure 5.2.2.

It is necessary to check that the proposed definition is unambiguous since two distinct curves γ and δ may have the same derivative at 0. We want to be sure that when this occurs the derivatives of the curves $f\circ\gamma$ and $f\circ\delta$ will also be equal at 0, as illustrated in Figure 5.2.3.

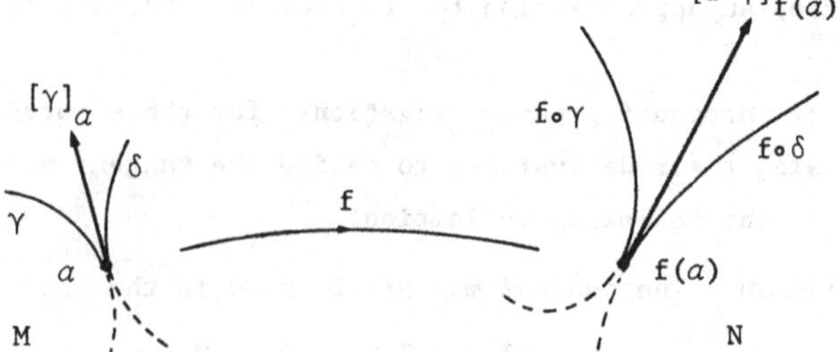

Figure 5.2.3. Tangency preserved.

5.2.2. Proposition. *For all differentiable parametrized curves* γ *and* δ *in* M *at* a

$$[\gamma]_a = [\delta]_a \Rightarrow [f\circ\gamma]_{f(a)} = [f\circ\delta]_{f(a)} .$$

Proof. It is sufficient to show that

$$\gamma'(0) = \delta'(0) \Rightarrow (f\circ\gamma)'(0) = (f\circ\delta)'(0).$$

The chain rule, Theorem 1.1.4, cannot be applied directly on the right hand side of this implication because the domain of f need not be an open set in a normed vector space. To overcome this difficulty, we introduce a chart (U, ϕ) for R^n at a with the submanifold property for M and then write, at least on a neighbourhood of $0 \in R$,

$$f\circ\gamma = (f\circ\phi^{-1})\circ\phi\circ\gamma .$$

Now by Corollary 1.4.4 and the chain rule

$$\begin{aligned}(f\circ\gamma)'(0) &= D(f\circ\gamma)(0)(1)\\ &= D(f\circ\phi^{-1})(\phi(a))\circ D\phi(\gamma(0))\circ D\gamma(0)(1)\\ &= D(f\circ\phi^{-1})(\phi(a))\circ D\phi(a)(\gamma'(0)).\end{aligned}$$

This, together with a similar expression for $(f\circ\delta)'(0)$, establishes the desired implication. ∎

Inasmuch as the Fréchet derivative of a map at any point of its domain is linear, the above expression for $(f\circ\gamma)'(0)$ shows it to be a linear function of $\gamma'(0)$.

5.2.3. Theorem. *The tangent map of* f *at* a

$$T_a f : T_a M \to T_{f(a)} N$$

is linear. ∎

With the aid of the tangent map, the chain rule for maps between manifolds may now be expressed in the following elegant way.

5.2.4. Theorem. (Chain Rule). *Let* $g: L \to M$ *and* $f: M \to N$ *be differentiable where* L, M *and* N *are submanifolds of Euclidean spaces. The composite* $f\circ g: L \to N$ *is then differentiable and*

$$T(f\circ g) = Tf\circ Tg .$$

If $a \in M$ *then we may also write*

$$T_a(f\circ g) = T_{g(a)}f\circ T_a g.$$

Proof. Proving the differentiability of the composite was set as Exercise 4.1.1.

To complete the proof of the theorem, let γ be a differentiable parametrized curve in L at a so that $g\circ\gamma$ is a differentiable parametrized curve in M at $g(a)$. From Definition 5.2.1 of the tangent map

$$\begin{aligned} T(f\circ g)([\gamma]_a) &= [(f\circ g)\circ\gamma]_{f(g(a))} \\ &= [f\circ(g\circ\gamma)]_{f(g(a))} \\ &= Tf([g\circ\gamma]_{g(a)}) \\ &= Tf(Tg([\gamma]_a)) \\ &= Tf\circ Tg([\gamma]_a). \quad \blacksquare \end{aligned}$$

EXERCISES 5.2.

1. *Show that* $T\mathrm{id}_M = \mathrm{id}_{TM}$.

2. *Clearly for each* C^r $(r \geq 1)$ *chart* (U,ϕ) *of the manifold* M $\phi\circ\phi^{-1} = \mathrm{id}_{\phi(U)}$ *and* $\phi^{-1}\circ\phi = \mathrm{id}_U$. *Use the chain rule and question 1 above to deduce that* $T\phi\colon TU \to T(\phi(U))$ *has both a left and a right inverse and that*
$(T\phi)^{-1} = T(\phi^{-1})$ *while* $T\phi(TU) = T(\phi(U))$.

3. *Let* $f\colon R^m \to R^n$ *and let* M *and* N *be submanifolds of* R^m *and* R^n *respectively, such that* $f(M) \subseteq N$. *Prove that if* f *is differentiable then so is its restriction* $f|M : M \to N$ *and that*
$$T(f|M) = Tf|TM.$$

5.3. TANGENT SPACES VIA IMPLICIT FUNCTIONS

We now turn to submanifolds of R^{n+m} which have the form

$$M = \{x\colon f(x) = 0\}$$

where $f\colon R^{n+m} \to R^m$ is a differentiable function for which $Df(a)$ has full rank m for each $a \in M$. For such submanifolds there is a very easy way to find the tangent space: just linearize the function f defining the submanifold near the appropriate point. This is the intuitive content of the following theorem.

5.3.1. Theorem. *The tangent space to the above submanifold at* $a \in M$ *is given by*

$$T_aM = \{(a,h): h \in \mathbb{R}^{n+m} \text{ with } Df(a)(h) = 0\}.$$

Proof. We will deal only with the case in which the partial derivative $\partial_2 f(a): \mathbb{R}^m \to \mathbb{R}^m$ has full rank m, leaving it as an exercise to derive the result in the remaining case.

Suppose first that $(a,h) \in T_aM$ so that $h = \gamma'(0)$ for some curve γ in M based at a. From the definition of M, $f \circ \gamma = 0$ and hence by the chain rule

$$Df(a)(\gamma'(0)) = 0 .$$

Thus (a,h) is a member of the set described in the theorem.

Conversely, suppose

$$Df(a)(h_1,h_2) = 0 \tag{1}$$

where $h = (h_1,h_2)$ with $h_1 \in \mathbb{R}^n$ and $h_2 \in \mathbb{R}^m$. If we now construct a curve γ in M based at a such that $h = \gamma'(0)$, then it will follow that $(a,h) \in T_aM$ so the theorem will be proved.

To this end note that since $\partial_2 f(a)$ has full rank, the implicit function theorem, Theorem 3.2.1, shows that there is a differentiable function g such that

$$f(x) = 0 \Longleftrightarrow x_2 = g(x_1) \tag{2}$$

for all $x = (x_1,x_2)$ in a neighbourhood of a. We define the curve $\gamma = (\gamma_1,\gamma_2)$ by putting

$$\gamma_1(t) = a_1 + th_1, \quad \gamma_2(t) = g(\gamma_1(t)) \tag{3}$$

for t sufficiently small, where $a = (a_1,a_2)$. Since $a_2 = g(a_1)$ it follows from (3) that $\gamma(0) = a$. From (2) and (3) it follows that

$$f(\gamma_1(t),\gamma_2(t)) = f(\gamma_1(t), g(\gamma_1(t))) = 0$$

so by the chain rule

$$Df(a)(\gamma_1'(0), \gamma_2'(0)) = 0. \tag{4}$$

Since, however, $\partial_2 f(a)$ has full rank, it follows from (1), (4) and $h_1 = \gamma_1'(0)$ that $h_2 = \gamma_2'(0)$; hence $h = \gamma'(0)$ as required. ■

EXERCISES 5.3.

1. *Use Theorem 5.3.1 to show that the tangent space to the unit sphere*

$$S^{n-1} = \{x \in R^n : x \cdot x = 1\}$$

at a point $a \in S^{n-1}$ *is*

$$T_a S^{n-1} = \{(a,h) : h \in R^n \text{ with } a \cdot h = 0\} .$$

2. *Use Theorem 5.3.1 to show that the tangent space to* $O(n)$, *defined in Exercise 3.3.2, at the identity matrix* I *is*

$$T_I O(n) = \{(I,H) : H \text{ is a skew-symmetric matrix of order } n\} .$$

3. *Use the ideas contained in Exercise 3.3.4 to extend the proof of Theorem 3.3.1 to the case in which* $\partial_2 f(a)$ *does not necessarily have full rank.*

6. TANGENT BUNDLES AS MANIFOLDS

In this chapter we show that the tangent bundle TM *of a manifold* M *is a manifold in its own right. Furthermore it has a differentiable structure that is induced naturally by the differentiable structure of* M. *These ideas can then be extended to form* T(TM) *which is the natural setting for second order differential equations which are fundamental in mechanics and other applications.*

6.1. CHARTS FOR TM.

Recall that the tangent bundle TM of a submanifold M of R^n is defined by

$$TM = \bigcup_{a \in M} T_aM \quad .$$

This is a subset of TR^n and hence may be regarded as a subset of $R^n \times R^n$.

For example, the tangent bundle TS^1 is the collection of all the tangent spaces to the unit circle. Although it is strictly a subset of R^4, it is nevertheless helpful to vizualize it in the plane by means of tangent lines attached to S^1, as in Figure 6.1.1.

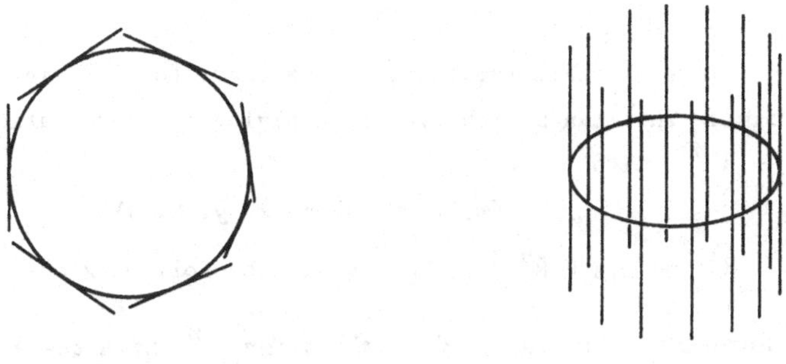

Figure 6.1.1. A tangent bundle TS^1

It is even easier to vizualize TS^1 if we give each tangent line a rotation perpendicular to the plane of the circle to form a cylinder, as illustrated in Figure 6.1.1. This latter interpretation is given formal justification in Exercise 6.1.1. Meanwhile the following results show how submanifold charts for M lead, via differentiation, to submanifold charts for TM.

6.1.1. Lemma. *Let* M *be a* $C^r (r \geq 2)$ *submanifold of* R^n *of dimension* k. *If* (U, ϕ) *is a chart for* R^n *with the submanifold property for* M, *then* $(TU, T\phi)$ *is a chart for* R^{2n} *such that*

$$T\phi(TU \cap TM) = T(\phi(U)) \cap (R^k \times \{0\} \times R^k \times \{0\}).$$

Proof. Let (U, ϕ) be as in the hypothesis of the lemma. By Exercise 5.2.2, the map $T\phi\colon TU \to T\phi(TU)$ is one-to-one and onto the set

$$T\phi(TU) = T(\phi(U)) = \phi(U) \times R^n,$$

which is open in R^{2n}. Hence $(TU, T\phi)$ is a chart for R^{2n}. Finally,

$$\begin{aligned} T\phi(TU \cap TM) &= T\phi(T(U \cap M)) && \text{by Exercise 5.1.5.} \\ &= T(\phi(U \cap M)) && \text{by Exercise 5.2.2.} \\ &= T(\phi(U) \cap (R^k \times \{0\})) && \text{by Definition 3.1.1.} \\ &= (\phi(U) \cap (R^k \times \{0\})) \cap (R^k \times \{0\} \times R^k \times \{0\}) && \text{by Definition 1.2.3.} \\ &= T\phi(U) \cap (R^k \times \{0\} \times R^k \times \{0\}) && \text{by Definition 1.2.3.} \end{aligned}$$ ∎

Thus all that remains to get a chart for TM is to shift zeros to the right. This can be achieved by applying the permutation $P : R^{2n} \to R^{2n}$ with

$$P(x, w, y, z) = (x, y, w, z)$$

for $x,y \in R^k$ and $w,z \in R^{n-k}$. This gives the following corollaries.

6.1.2. Corollary. *If* (U,ϕ) *is a chart for* R^n *with the submanifold property for* M *then* $(TU, P\circ(T\phi))$ *is a chart for* R^{2n} *with the submanifold property for* TM. ∎

6.1.3. Corollary. *If* M *is a* $\mathbb{C}^r$ $(r \geq 2)$ *submanifold of* R^n *then* TM *is a* $\mathbb{C}^{r-1}$ *submanifold of* R^{2n}. *The dimension of* TM *is twice the dimension of* M. ■

The following example provides an illustration of Corollary 6.1.2.

6.1.4. Example. A chart $(\mathrm{U},(\theta,\ \mathrm{r}-\underline{1}))$ on R^2 is defined by

$$\mathrm{U} = \{(a,b) \in \mathrm{R}^2 : a > 0\}$$

$$\theta(a,b) = \arctan\left(\frac{b}{a}\right)$$

$$(\mathrm{r}-\underline{1})\ (a,b) = \sqrt{a^2 + b^2} - 1\ .$$

This chart has the submanifold property for S^1 since, on $\mathrm{U} \cap \mathrm{S}^1$, $a^2 + b^2 = 1$ so $(\theta,\ \mathrm{r}-\underline{1})(a,b) = \left(\arctan\left(\frac{b}{a}\right),\ 0\right)$.

We now compute $\mathrm{T}(\theta,\ \mathrm{r}-\underline{1})$ and show that, to within a permutation, this is a chart for TR^2 with the submanifold property for TS^1. Let $(a,b) \in \mathrm{U}$ and $(h,k) \in \mathrm{R}^2$. Then

$$\begin{aligned} \mathrm{D}\theta(a,b)(h,k) &= \mathrm{D}\arctan\left(\frac{b}{a}\right)\circ \mathrm{D}\left(\frac{\mathrm{id}_2}{\mathrm{id}_1}\right)(a,b)(h,k) \\ &= \frac{\mathrm{id}}{1+\left(\frac{b}{a}\right)^2}\left(-\frac{b\ \mathrm{id}_1}{a^2} + \frac{\mathrm{id}_2}{a}\right)(h,k) \\ &= \frac{-bh + ak}{a^2 + b^2} \end{aligned}$$

and

$$\begin{aligned} \mathrm{D}(\mathrm{r}-\underline{1})(a,b)(h,k) &= \mathrm{D}(\mathrm{id}^{\frac{1}{2}}\circ(\mathrm{id}_1{}^2+\mathrm{id}_2{}^2)-\underline{1})(a,b)(h,k) \\ &= \mathrm{Did}^{\frac{1}{2}}(a^2+b^2)\circ \mathrm{D}(\mathrm{id}_1{}^2+\mathrm{id}_2{}^2)(a,b)(h,k) \\ &= \frac{ah + bk}{\sqrt{a^2 + b^2}}\ . \end{aligned}$$

Putting this together and identifying TR^2 with R^4 we obtain

$$\mathrm{T}(\theta,\ \mathrm{r}-\underline{1})(a,b)(h,k) = \left(\arctan\left(\frac{b}{a}\right),\ \sqrt{a^2+b^2}-1,\ \frac{-bh + ak}{a^2 + b^2},\ \frac{ah + bk}{\sqrt{a^2 + b^2}}\right).$$

Restricting this to TS^1 gives

$$\mathrm{T}(\theta,\ \mathrm{r}-\underline{1})\Big|_{\mathrm{TS}^1}(a,b)(h,k) = \left(\arctan\left(\frac{b}{a}\right),\ 0,\ \frac{-bh + ak}{a^2 + b^2},\ 0\right)$$

since by Exercise 5.1.1. $ah + bk = 0$ on S^1. Thus $(U \times R^2, T(\theta, r-\underline{1}))$ has the submanifold property for TS^1 as required. ■

6.1.5. Theorem. *If* (U, ϕ) *is a* C^r $(r \geq 2)$ *chart for a submanifold* M *of* R^n *then* $(TU, T\phi)$ *is a* C^{r-1} *chart for* TM.

Proof. Follows from Corollaries 6.1.2 and 6.1.3. ■

EXERCISES 6.1.

1. *Show that the map, from the tangent bundle of the unit circle to the cylinder in* R^3,

$$\Phi: TS^1 \to \{(a,b,c) \in R^3 : a^2 + b^2 = 1\}$$

given by

$$\Phi((a,b),(h,k)) = (a,b, -bh + ak)$$

is a diffeomorphism.

(Hint: refer to Example 6.1.4. and then construct charts for TS^1 *and the cylinder with respect to which the local representative of* Φ *is the identity mapping).*

2. *Let* $U = R^2 \backslash \{(0,b) : b \geq 0\}$ *and*

$$\psi_N(a,b) = \frac{a}{\sqrt{a^2+b^2} - b}$$

$$(r-\underline{1})(a,b) = \sqrt{a^2 + b^2} - 1.$$

(See Figure 6.1.2.)

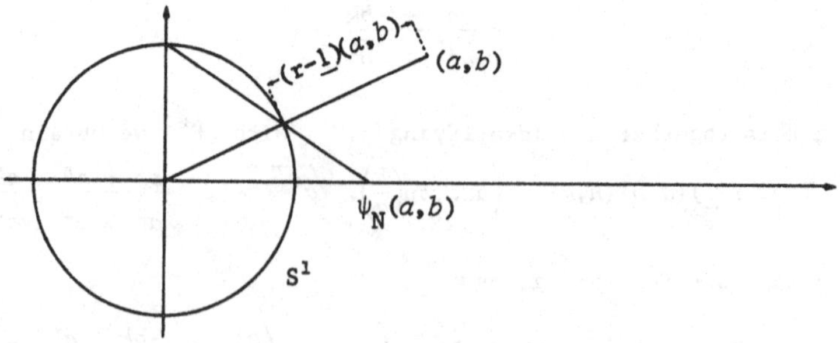

Figure 6.1.4. Extended stereographic chart.

(a) *Show that* $(U, (\psi_N, r-1))$ *is a chart for* R^2 *with the submanifold property for* S^1.

(b) *Use an argument similar to that of example 6.1.4 to show that* $(TU, T(\psi_N, r-\underline{1}))$ *is a chart for* TR^2 *with the submanifold property for* TS^1.

6.2. PARALLELIZABILITY

In the previous section we showed that $TS^1 = S^1 \times R$ and made the observation that $TR^n = R^n \times R^n$. It therefore seems natural to conjecture that in general $TM = M \times R^k$. This conjecture is false.

6.2.1. Definitition. Let M be C^r $(r \geqslant 2)$ k-dimensional submanifold of R^n. If there exists a diffeomorphism

$$\Phi : TM \longrightarrow M \times R^k$$

which takes each tangent space T_aM by a linear isomorphism to $\{a\} \times R^k$ then M is called *parallelizable.* ■

An important example of a manifold that is not parallelizable and which arises in the study of the spherical pendulum is S^2.

6.2.2. Theorem. S^{2n} *is not parallelizable.* The proof of this theorem involves the use of the "Hairy Ball Theorem" (see Hirsch(1976)) and for a sketch of the proof itself see Chillingworth(1976). ■

We do, however, have "local" parallelizability as stated in the following lemma.

6.2.3. Lemma. *Let* M *be a* C^r $(r \geqslant 2)$ *k-dimensional submanifold of* R^n *and let* (U, ϕ) *be an admissible chart for* M. *Then* $TU = U \times R^k$.

Proof. See exercises. ■

We quote a deep result, mentioned in Chillingworth(1976), whose proof lies outside the scope of this book.

6.2.4. Theorem. $TS^n = S^n \times R^n$ *only in the cases* $n = 0, 1, 3, 7$. ■

EXERCISES 6.2.

1. *Prove Lemma 6.2.3. by quoting the appropriate result from Section 6.1.*

2. *Suppose* M *is a Möbius strip (See Figure 6.2.1). Is* M *parallelizable? Give géometric reasons for your answer.*

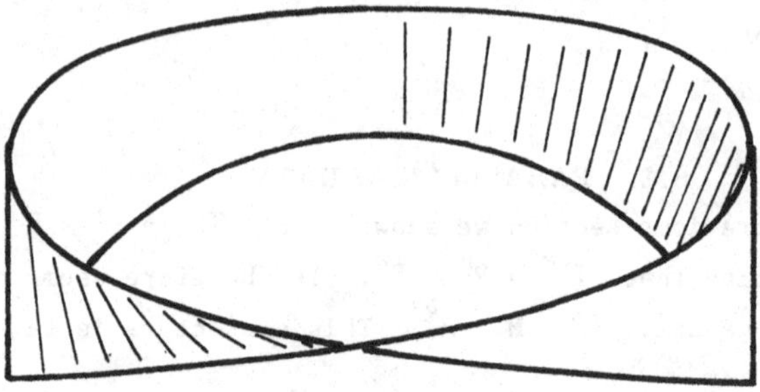

Figure 6.2.1. A Möbius Strip

6.3. TANGENT MAPS AND SMOOTHNESS

Corollary 6.1.3. enables use to give a tangent bundle the structure of a manifold. Put simply if (U, ϕ) is a chart for M then $(TU, T\phi)$ is a chart for TM. There are clearly other admissible charts for TM which are not obtained in this way but they are not of interest to us since they may not preserve the linearity of the tangent spaces and hence may not preserve the linearity of local representatives of tangent maps.

6.3.1. Definition. Let $\{(U_i, \phi_i)\colon i \in I\}$ be an atlas for the $\mathbb{C}^r$ $(r \geq 2)$ submanifold M of R^n. Then the collection of charts for TM given by $\{(TU_i, T\phi_i)\colon i \in I\}$ is called the *natural atlas* for TM. ■

It is now meaningful to discuss differentiability of maps which have tangent bundles as their domains. An important example of such a map is given by the following definition.

6.3.2. Definition. Let M be a $\mathbb{C}^r$ $(r \geq 2)$ submanifold of R^n. The *natural projection* on the tangent bundle of M is the map $\tau_M\colon TM \to M$ with

$$\tau_M(a,h) = a$$

for each $(a,h) \in T_aM$. ■

6.3.3. Theorem. *The natural projection* τ_M *is of class* $\mathfrak{C}^\infty$.

Proof. We check the conditions in Definition 4.1.2. Let $(a,h) \in TM$ and let (U, ϕ) be a $\mathfrak{C}^r$ chart for M. By Theorem 6.1.5, $(TU, T\phi)$ is a chart for TM and $\tau_M(TU) \subseteq U$.

The local representative of τ_M relative to these charts is the map

$$\phi \circ \tau_M \circ (T\phi)^{-1} \quad .$$

From Figure 6.3.1, which is validated by the definition of the tangent map $T\phi$, it is clear that this map just the natural projection $\Pi: R^{2n} \to R^n$, which is $\mathfrak{C}^\infty$. ■

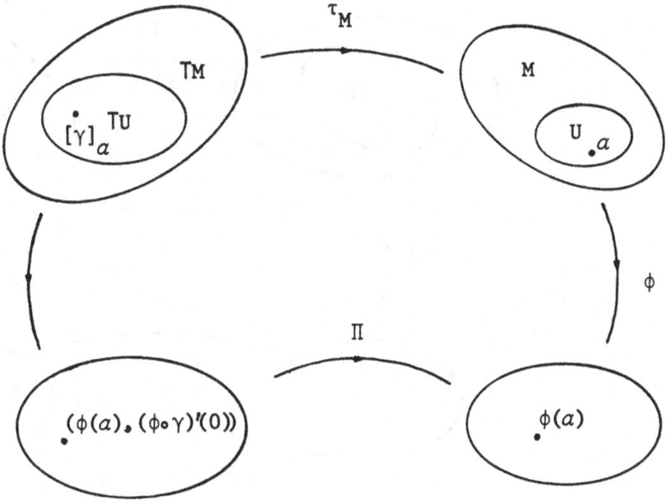

Figure 6.3.1. Local representative of τ_M.

Since formation of a tangent map is essentially a process of differentiation, it is not surprising that it may lead to a loss of one degree of differentiability.

6.3.3. Theorem. *Let* M *and* N *be* $\mathfrak{C}^r$ $(r \geqslant 2)$ *submanifolds of* R^m *and* R^n *respectively. If* $f : M \to N$ *is of class* $\mathfrak{C}^r$ *then* $Tf: TM \to TN$ *is of class* $\mathfrak{C}^{r-1}$.

Proof. This again involves straightforward checking of the conditions in Definition 4.1.2. ■

EXERCISES 6.3.

1. *Let* $f : M \to N$ *be as in Theorem 6.3.3. Show that the local representative of* Tf *satisfies*

$$Tf_{T\phi\, T\psi} = T\,f_{\phi\psi}$$

where (U, ϕ) *is a chart for* M, (V, ψ) *is a chart for* N *and* $f(U) \subseteq V$. *See Figure 6.3.2.*

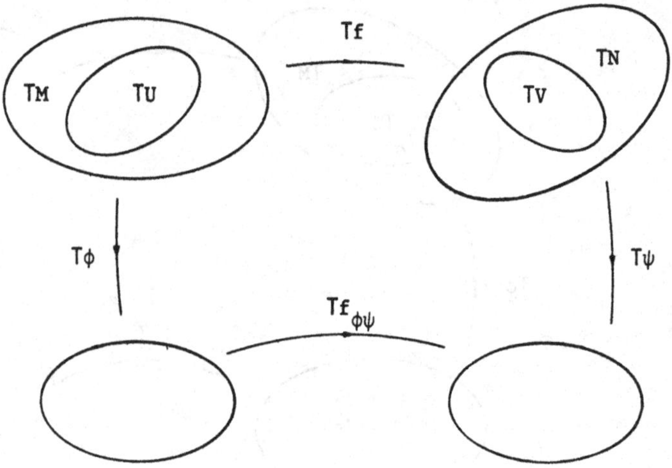

Figure 6.3.2. Local representative of Tf.

2. *Use question 1 to fill in the details of the proof of Theorem 6.3.3.*

6.4. DOUBLE TANGENTS.

In Section 6.1 we showed that if M was a manifold then TM was also a manifold with a differentiable structure which is induced naturally by that of M. It is possible to repeat this process and obtain an atlas for $T(TM)$. We will denote by T^2M the manifold $T(TM)$.

6.4.1. Lemma. *Let* M *be a* $\mathcal{C}^r (r \geq 3)$ *submanifold of* $\mathbb{R}^n$ *of dimension* k *and suppose* (U, ϕ) *is an admissible chart for* M. *Then* $(T^2U, T^2\phi)$ *is a chart for* T^2M. *Furthermore* T^2M *is a submanifold of* $(\mathbb{R}^n)^4$ *of dimension* $4k$.

Proof. Follows from Lemma 6.1.1, Corollary 6.1.2. and Corollary 6.1.3. ■

Various other results follow from those of previous sections for example:

$$T^2(f \circ g) = T^2f \circ T^2g$$

and
$$T^2f_{T^2\phi, T^2\psi} = T^2f_{\phi\psi}$$

when each of the above expressions is properly defined. We will not fill in the details but the following lemma gives a little insight into the structure of double tangent maps.

6.4.2. Lemma. *Let* $g: \mathbb{R}^n \to \mathbb{R}^m$ *be twice differentiable. Then*

$$T^2g: (\mathbb{R}^n)^4 \to (\mathbb{R}^m)^4 \quad \textit{and}$$

$$T^2g(a,h,k,\ell) = (g(a), Dg(a)(h), Dg(a)(k), D^2g(a)(h,k) + Dg(a)(\ell)).$$

Proof. Follows from Definitions 1.1.3 and 1.2.4. ■

EXERCISES 6.4.

1. *Fill in the details of the proof of Lemma 6.4.2.*

2. *Notice that* $\tau_M: TM \to M$ *and* $\tau_M(a,h) = a$. *Thus* $T\tau_M : T^2M \to TM$.

   (a) *Suppose* $(a,h,k,\ell) \in T^2M$. *Find* $T\tau_M(a,h,k,\ell)$.

   (b) *Show that* $\tau_M \circ T\tau_M = \tau_M \circ \tau_{TM}$.

   (c) *Figure 6.3.1. shows the local representative of* τ_M *with respect to natural charts. Find the local representatives of* $T\tau_M$ *and* τ_{TM} *with respect to suitable natural charts.*

   (d) *For what subset of* T^2M *is* $\tau_{TM} = T\tau_M$?

7. PARTIAL DERIVATIVES

The tangent bundle of a differentiable manifold provides a natural setting for the study of particle motion. Before deriving Lagrange's equations of motion we establish some technical results on partial differentiation with respect to charts. The resulting formulas have a classical look but are here given a precise and useful meaning.

7.1. CURVES IN TQ

To specify the state of a particle moving on a manifold Q at any given time one specifies both the position and the velocity of the particle. Hence it is natural to regard the particle as tracing out a curve in TQ rather than in Q. The motion of the particle is thus described by a parametrized curve

$$c : I \to TQ$$

where I is an open interval in R. We may write c in terms of its components as (c_1, c_2) where c_1 is the projection of c onto Q,

$$\tau_Q \circ c : I \to Q,$$

which traces out only the position of the particle, as in figure 7.1.1.

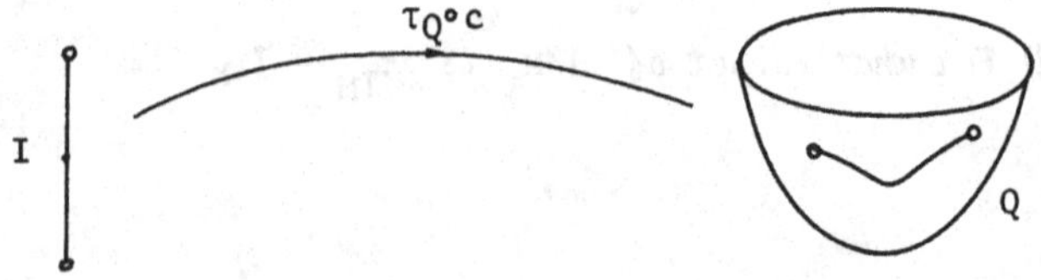

Figure 7.1.1. The position of a particle.

Intuitively

$$c : I \longrightarrow TQ$$

can be thought of as the parametrized curve $\tau_Q \circ c$ together with an attached "arrow" at each point of $\tau_Q \circ c(I)$, as in Figure 7.1.2.

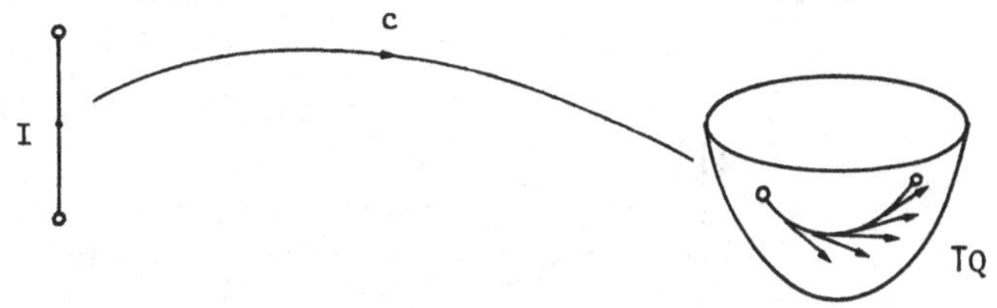

Figure 7.1.2. The position and velocity.

In this interpretation $c(t)$ consists of $\tau_Q \circ c(t)$ (the position of the particle on Q) together with an "arrow" (the velocity of the particle). The arrow should point in the direction of motion of the particle, that is, it should be tangential to the image of $\tau_Q \circ c$ at the point in question and the length of the arrow should give the speed of the particle. These requirements may be expressed in terms of the components of c by the condition that

$$c_2(t) = c_1'(t)$$

and hence that

$$c(t) = (c_1(t), c_1'(t)).$$

This leads to the following definitions.

7.1.1. Definition. Let Q be a manifold and $c = (c_1, c_2): I \longrightarrow TQ$ a differentiable parametrized curve. We define

$$c_1^{\dot{}}(t) = (c_1(t), c_1'(t))$$

$$(= Tc_1(t,1)). \blacksquare$$

7.1.2. Definition. A parametrized curve $c: I \longrightarrow TQ$ is said to be *self-consistent* if $(\tau_Q \circ c)^{\dot{}} = c$. ■

7.1.3. **Example.** Let $c : R \to TR$ be given by $c(t) = (t^2, 2t)$ for each $t \in R$. Here

$$(\tau_R \circ c)(t) = t^2$$
$$(\tau_R \circ c)^{;}(t) = (t^2, 2t)$$

and hence

$$(\tau_R \circ c)^{;} = c$$

so that the parametrized curve c is self-consistent. The effect of c on

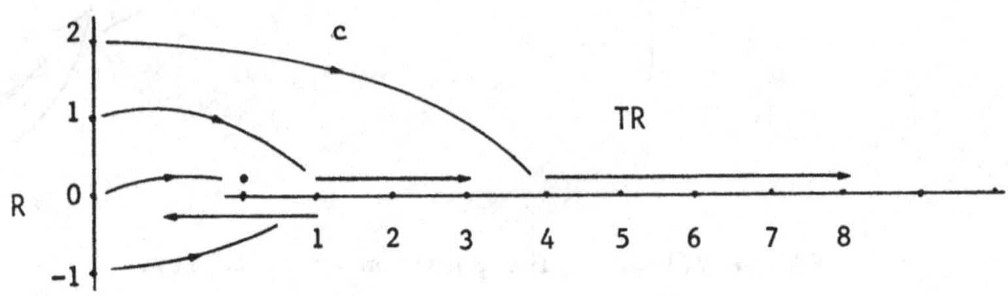

Figure 7.1.3.

some elements of R is indicated in Figure 7.1.3 (c.f. Figure 7.1.2). Identifying TR with R^2 gives the diagram shown in Figure 7.1.4. ■

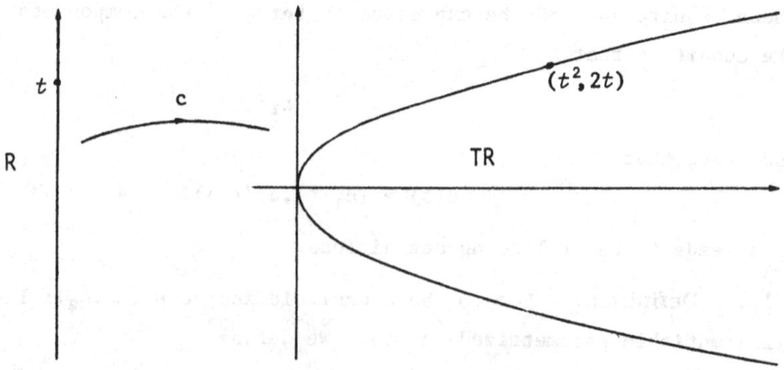

Figure 7.1.4.

EXERCISES 7.1.

1. *Show that* $c : R \to TR$ *given by* $c(t) = (\sin(t), \cos(t))$ *is a self-consistent path in* TR *and sketch its trajectory in* TR *in the same manner as Figures 7.1.3 and 7.1.4.*

2. *Show that* $c : R \to TR^2$ *given by*

$$c(t) = ((\sin(t), \cos(t)), (\cos(t), -\sin(t))$$

is a self-consistent path in TR^2. *Sketch an "arrow diagram" for this path in* R^2.

7.2. TRADITIONAL NOTATION

The traditional notation for partial derivatives is often used in an ambiguous way. We give a rigorous definition which is valid in the context of partial differentiation with respect to the component functions of a chart.

To this end let Q be a manifold of dimension k and let $f : Q \to R$ be differentiable. If (U, ϕ) is a chart for Q then the local representative f_ϕ of f satisfies

$$f = f_\phi \circ \phi$$

as shown in Figure 7.2.1. Now f_ϕ maps an open set in Euclidean space into R, hence the Definition 1.4.1 of $f_\phi{}^{/i}$ is applicable.

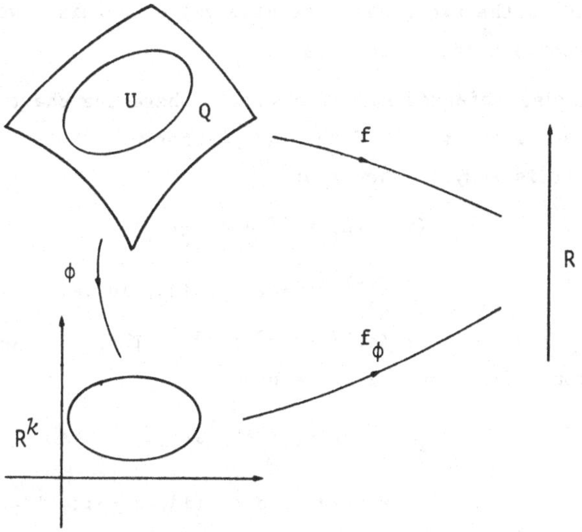

Figure 7.2.1. Local representative of real-valued f.

7.2.1. Definition. Let (U, ϕ) be a chart for the manifold Q and, in terms of components, let

$$\phi = (\phi_1, \phi_2, \ldots, \phi_k).$$

The *partial derivative of* f *with respect to* ϕ_i, denoted by

$$\frac{\partial f}{\partial \phi_i} : U \to R$$

is defined by putting

$$\frac{\partial f}{\partial \phi_i} = f_\phi{}^{/i} \circ \phi$$

where $f_\phi = (f \circ \phi^{-1})$ is the local representative of f. ■

Notice that in the above definition all of the components of the chart function ϕ are involved. This is often emphasised in the literature by writing

$$\frac{\partial f}{\partial \phi_i} \quad \text{as} \quad \left(\frac{\partial f}{\partial \phi_i}\right)_{(\phi_1, \ldots, \phi_{i-1}, \phi_{i+1}, \ldots, \phi_k)}.$$

A good account of the confusion which can arise when this point is overlooked is given in Munroe(1963) Chapter 5.

In the classical approach the "coordinate functions" ϕ_i were called "real-variables". We can still think of $\frac{\partial f}{\partial \phi_i}$ as being the derivative of f with respect to ϕ_i keeping the other ϕ_j's fixed. The traditional rules for calculating with partial derivatives, some of which are contained in the exercises, are also valid in this newer context. The following example illustrates this.

7.2.2. Example. Refer to Exercise 2.1.5 where the chart $(U,(r,\theta))$ for R^2 was defined with $\{(a,b) \in R^2 : a > 0\}$ for U.

It is easy to show that

$$(r,\theta)(U) = R^+ \times (-\tfrac{\pi}{2}, \tfrac{\pi}{2})$$

$$(r,\theta)^{-1} = (id_1 . \cos \circ id_2,\ id_1 . \sin \circ id_2) \quad .$$

Now suppose $f: R^2 \to R \ : (a,b) \mapsto a^2 + b^2$. Thus $f = id_1{}^2 + id_2{}^2$ ($= r^2$) so on U from Definition 7.2.1. we have

$$\frac{\partial f}{\partial r} = (f \circ (r,\theta)^{-1})^{/1} \circ (r,\theta)$$

$$= ((id_1{}^2 + id_2{}^2) \circ (id_1 . \cos \circ id_2 , id_1 . \sin \circ id_2))^{/1} \circ (r,\theta)$$

$$= (\mathrm{id}_1{}^2)^{/1} \circ (r,\theta)$$

$$= 2\mathrm{id}_1 \circ (r,\theta)$$

$$= 2r.$$

$$\text{and } \frac{\partial f}{\partial \theta} = (\mathrm{id}_1{}^2)^{/2} \circ (r,\theta)$$

$$= \underline{0}. \quad \blacksquare$$

It will be convenient to define the partial derivative of a function g which maps into R^n as a column vector.

7.2.3. Definition. Let Q be a manifold of dimension k with chart (U, ϕ) and let g be a $\mathbf{C}^r$ $(r \geqslant 1)$ map

$$g\colon Q \to \mathsf{R}^n .$$

We define *the derivative of* g *with respect to* ϕ_i as the column vector

$$\frac{\partial g}{\partial \phi_i} = \begin{pmatrix} (g_1 \circ \phi^{-1})^{/i} \circ \phi \\ \vdots \\ (g_n \circ \phi^{-1})^{/i} \circ \phi \end{pmatrix}. \quad \blacksquare$$

Notice that if $h\colon \mathsf{R}^n \to \mathsf{R}$ is differentiable then $h'\colon \mathsf{R}^n \to \mathsf{R}$ can be thought of as the row vector of functions $(h_1{}', h_2{}', \ldots, h_n{}')$. The following lemma should be read with this in mind.

7.2.4. Lemma. (Semi-classical chain rule).

Suppose $g\colon Q \to \mathsf{R}^n$ *and* $h\colon \mathsf{R}^n \to \mathsf{R}$ *are differentiable. If* (U,ϕ) *is a chart for the manifold* Q *then*

$$\frac{\partial h \circ g}{\partial \phi_i} = \sum_{j=1}^{n} h^{/j} \circ g \; \frac{\partial g_j}{\partial \phi_i} = h' \circ g \; \frac{\partial g}{\partial \phi_i} .$$

Proof.

$$\frac{\partial h \circ g}{\partial \phi_i} = (h \circ g \circ \phi^{-1})^{/i} \circ \phi \qquad \text{by Definition 7.2.1.}$$

$$= \sum_{j=1}^{n} (h^{/j} \circ g \circ \phi^{-1} (g \circ \phi^{-1})_j^{/i}) \circ \phi$$

by the chain rule, Theorem 1.4.5. ■

A further generalization of Definition 7.2.1. which involves the partial derivative of a real valued function with respect to all of a chart's component functions is as follows:

7.2.5. Definition. Suppose Q is a manifold of dimension k and $f\colon Q \to R$ is differentiable. For each chart (U, ψ) for Q we define the *partial derivative of* f *with respect to* ψ *as the row vector*

$$\frac{\partial f}{\partial \psi} = \left(\frac{\partial f}{\partial \psi_1}, \ldots, \frac{\partial f}{\partial \psi_k}\right). \ \blacksquare$$

We can now formulate a classical looking chain rule.

7.2.6. Theorem. (Classical chain rule). *Let* Q *be a k-dimensional manifold, and* $f\colon Q \to R$ *be differentiable. If* (U, ϕ) *and* (V, ψ) *are charts for* Q *with* $U \cap V \neq \square$, *then*

$$\frac{\partial f}{\partial \phi_i} = \frac{\partial f}{\partial \psi} \cdot \frac{\partial \psi}{\partial \phi_i} .$$

Proof.

$$\frac{\partial f}{\partial \phi_i} = (f\circ\phi^{-1})^{/i}\circ\phi \qquad \text{by Definition 7.2.1}$$

$$= ((f\circ\psi^{-1})\circ(\psi\circ\phi^{-1}))^{/i}\circ\phi$$

$$= \sum_{j=1}^{k}\left((f\circ\psi^{-1})^{/j}\circ(\psi\circ\phi^{-1})\ (\psi\circ\phi^{-1})_j^{/i}\right)\circ\phi \qquad \text{by the chain rule for partial derivatives Theorem 1.4.5.}$$

$$= \sum_{j=1}^{k}(f\circ\psi^{-1})^{/j}\circ\psi\ (\psi\circ\phi^{-1})_j^{/i}\circ\phi$$

$$= \sum_{j=1}^{k}\frac{\partial f}{\partial \psi_j}\ \frac{\partial \psi_j}{\partial \phi_i} \qquad \text{by Definition 7.2.1}$$

$$= \frac{\partial f}{\partial \psi} \cdot \frac{\partial \psi}{\partial \phi_i} . \ \blacksquare$$

The following definition generalizes Definitions 7.2.1 and 7.2.3.

7.2.7. Definition. Let (U, ϕ) and (V, ψ) where $U \cap V \neq \square$ be charts for the k-dimensional manifold Q. Then *the derivative of* ψ *with respect to* ϕ, denoted by $\frac{\partial \psi}{\partial \phi}$, is the $k \times k$ matrix whose i-jth component is $\frac{\partial \psi_i}{\partial \phi_j}$. $\blacksquare$

The above definition will be useful later as will be the following obvious Corollary to Theorem 7.2.6.

7.2.8. Corollary. $\dfrac{\partial f}{\partial \phi} = \dfrac{\partial f}{\partial \psi} \dfrac{\partial \psi}{\partial \phi}$. ∎

The classical notion of a "differential" has a modern analogue as shown by the following theorem.

7.2.9. Theorem. *If* (U, ϕ) *is a chart for* R^n *and* $f : U \to R$ *is differentiable, then*

$$Df = \sum_{i=1}^{n} \frac{\partial f}{\partial \phi_i} D\phi_i .$$

Proof. Let $a \in U$.

$$\begin{aligned} Df(a) &= D(f\circ\phi^{-1}\circ\phi)(a) \\ &= D(f\circ\phi^{-1})(\phi(a))\circ D\phi(a) && \text{by Theorem 1.1.4.} \\ &= \sum_{i=1}^{n} (f\circ\phi^{-1})^{/i}(\phi(a))(D\phi(a))_i && \text{by Theorem 1.4.3.} \\ &= \sum_{i=1}^{n} (\frac{\partial f}{\partial \phi_i} D\phi_i)\ (a) && \text{by Definition 7.2.1.} \end{aligned}$$ ∎

The following version of the chain rule expresses what is sometimes referred to in more traditional books as the rule for the "total derivative" of a function.

7.2.10 Theorem. *Let* (U, ϕ) *be a chart for a k-dimensional manifold* Q. *If* $\gamma : I \to Q$ *is a differentiable parametrized curve and* $f : Q \to R$ *is differentiable then*

$$(f\circ\gamma)' = \sum_{j=1}^{k} (\frac{\partial f}{\partial \phi_j}\circ\gamma)(\phi_j\circ\gamma)' .$$

Proof.

$$\begin{aligned} (f\circ\gamma)' &= (f\circ\phi^{-1}\circ\phi\circ\gamma)' \\ &= \sum_{j=1}^{k} (f\circ\phi^{-1})^{/j}\circ\phi\circ\gamma \quad (\phi\circ\gamma)_j' && \text{by Theorem 1.4.5.} \\ &= \sum_{j=1}^{k} (\frac{\partial f}{\partial \phi_j}\circ\gamma)(\phi_j\circ\gamma)' && \text{by Definition 7.2.1.} \end{aligned}$$ ∎

EXERCISES 7.2.

In these exercises (U,ϕ) *will denote a chart for a manifold* Q *and we let* $\phi = (\phi_1, \phi_2, \ldots, \phi_k)$.

1. *Show that for* $i,j \in \{1,2,\ldots,k\}$

$$\frac{\partial \phi_i}{\partial \phi_j} = \begin{cases} 1 & \text{if } i=j \\ 0 & \text{if } i \neq j. \end{cases}$$

2. *Let* $f:Q \to R$ *and* $g:Q \to R$ *be differentiable.*
Prove each of the following familiar looking rules:

$$\frac{\partial(f+g)}{\partial \phi_i} = \frac{\partial f}{\partial \phi_i} + \frac{\partial g}{\partial \phi_i}, \quad \frac{\partial(fg)}{\partial \phi_i} = f\frac{\partial g}{\partial \phi_i} + g\frac{\partial f}{\partial \phi_i}.$$

(*Assume the corresponding rules for* $(f+g)^{/i}$ *and* $(fg)^{/i}$).

3. *Let* (x,y) *denote the identity chart for* R^2, *that is, let* $(x,y)(a,b) = (a,b)$ *for each* $(a,b) \in R^2$. *Show that, for* $(U,(r,\theta))$ *as in Example 7.2.2,*

$$\frac{\partial r}{\partial x} = \frac{x}{\sqrt{x^2+y^2}}, \quad \frac{\partial r}{\partial y} = \frac{y}{\sqrt{x^2+y^2}}$$

$$\frac{\partial \theta}{\partial x} = \frac{-y}{x^2+y^2}, \quad \frac{\partial \theta}{\partial y} = \frac{x}{x^2+y^2}$$

Hence write out the matrix expression for $\frac{\partial(r,\theta)}{\partial(x,y)}$.

7.3 SPECIALIZATION TO TQ

The notation introduced in the previous section is applicable to functions defined on any manifold hence, in particular, it is applicable to functions defined on the tangent bundle TQ of a submanifold Q of R^n. In this more specialized context additional results on partial differentiation can be obtained. We assume throughout that Q has dimension k.

The dot notation is used ambiguously in the traditional books on mechanics. Here we assign a rigorous meaning to one of its two possible interpretations (and ignore the other in which it is regarded as differentiation with respect to time). A key role will be played by the idea of a self-consistent path, introduced in Section 7.1.

7.3.1. Definition. For a differentiable map $f : Q \to R$ we define $\dot{f} : TQ \to R$ by putting, for $[\gamma]_a \in TQ$,

$$\dot{f}([\gamma]_a) = \pi_2 \circ Tf([\gamma]_a) = (f \circ \gamma)'(0)$$

where $\pi_2 \colon R \times R \longrightarrow R$ denotes projection onto the second factor. ■

The maps involved in the definition of $\dot{f}$ are illustrated in Figure 7.3.1.

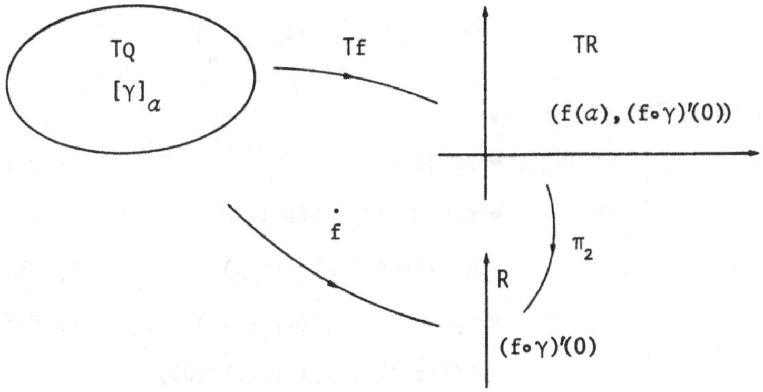

Figure 7.3.1. The $\dot{f}$ map

Now let (U, ϕ) be a chart for a manifold Q. Since each component ϕ_i maps U into R we may apply the above definition to obtain a map

$$\dot{\phi}_i : TU \longrightarrow R$$

such that for $[\gamma]_a \in TU$

$$\dot{\phi}_i([\gamma]_a) = (\phi_i \circ \gamma)'(0).$$

Hence, by Definition 5.2.1, we may write

$$T\phi = (\phi \circ \tau_Q, \dot{\phi}) .$$

But since, by Definition 6.3.1, $(TU, T\phi)$ is a chart on the tangent bundle TQ, we may partially differentiate with respect to the components of the map $(\phi \circ \tau_Q, \dot{\phi})$. To avoid repeated use of cumbersome symbolism, we introduce the following definition.

7.3.2. Definition. Let $g \colon TQ \longrightarrow R$ be differentiable and let $(TU, (\phi \circ \tau_Q, \dot{\phi}))$ be a natural chart on TQ. For $1 \leqslant i \leqslant k$, we define

$$\frac{\partial g}{\partial \phi_i} : TU \longrightarrow R$$

to be the map $\dfrac{\partial g}{\partial(\phi_i\circ\tau_Q)}$. ∎

Note that the earlier Definition 7.2.1 is not applicable here because g and ϕ do not have a common domain.

7.3.3. Theorem. *For* f *as in Definition 7.3.1 and* ϕ *as above,*

$$\dot{f} = \sum_{i=1}^{k}\left(\frac{\partial f}{\partial\phi_i}\circ\tau_Q\right)\dot{\phi}_i .$$

Proof. For $[\gamma]_a \in TQ$,

$$\begin{aligned}
\dot{f}([\gamma]_a) &= \pi_2\circ Tf([\gamma]_a) && \text{by Definition 7.3.1.}\\
&= \pi_2\circ T(f\circ\phi^{-1}\circ\phi)([\gamma]_a) &&\\
&= \pi_2\circ T(f\circ\phi^{-1})\circ T\phi([\gamma]_a) && \text{by Theorem 5.2.4.}\\
&= \pi_2\circ T(f\circ\phi^{-1})(\phi(a),(\phi\circ\gamma)'(0)) && \text{by Definition 5.2.1.}\\
&= D(f\circ\phi^{-1})(\phi(a))((\phi\circ\gamma)'(0)) &&\\
&= \sum_{i=1}^{k}(f\circ\phi^{-1})^{/i}(\phi(a))((\phi\circ\gamma)'(0))_i && \text{by Theorem 1.4.3.}\\
&= \sum_{i=1}^{k}(f\circ\phi^{-1})^{/i}(\phi(a))(\phi_i\circ\gamma)'(0) &&\\
&= \sum_{i=1}^{k}\frac{\partial f}{\partial\phi_i}(a)\dot{\phi}_i([\gamma]_a) && \text{by Definition 7.2.1.} \blacksquare
\end{aligned}$$

7.3.4. Corollary. (Cancellation of dots rule)

$$\frac{\partial\dot{f}}{\partial\dot{\phi}_i} = \frac{\partial f}{\partial\phi_i}\circ\tau_Q . \quad\blacksquare$$

7.3.5. Corollary. $\dfrac{\partial\dot{f}}{\partial\phi_i} = \left(\displaystyle\sum_{j=1}^{k}\frac{\partial^2 f}{\partial\phi_i\partial\phi_j}\circ\tau_Q\right)\dot{\phi}_i$. ∎

7.3.6. Theorem. (Interchange of dot and dash 1). *If* $c : I \to TQ$ *is a self-consistent path in* TQ *then, for* $f: Q \to R$ *differentiable,*

$$\dot{f}\circ c = (f\circ\tau_Q\circ c)' .$$

Proof.

$$\begin{aligned}
\dot{f}(c(t)) &= \pi_2\circ Tf(c(t)) && \text{by Definition 7.3.1.}\\
&= \pi_2\circ Tf\ (T\tau_Q\circ c(t,1)) && \text{by Definition 7.1.1.}\\
&= \pi_2\circ T(f\circ\tau_Q\circ c)(t,1) && \text{by the chain rule}\\
&= (f\circ\tau_Q\circ c)'(t). \quad\blacksquare
\end{aligned}$$

7.3.7. Corollary. *For* $c: I \to TQ$ *a self-consistent path,* (U,ϕ) *a chart for* Q,

$$(f\circ\tau_Q\circ c)' = \sum_{j=1}^{k}\left[\frac{\partial f}{\partial\phi_j}\circ\tau_Q\circ c\right](\dot\phi_j\circ c).$$

Proof. See exercises. ■

7.3.8. Corollary. (Interchange of dot and dash 2).

$$\frac{\partial\dot f}{\partial\phi_j}\circ c = \left[\frac{\partial f}{\partial\phi_j}\circ\tau_Q\circ c\right]'$$

for each self-consistent path $c : I \to TQ$.

Proof. See exercises. ■

7.3.9. Theorem. *Let* Q *be a submanifold of* R^n *and let* $(TU, T\phi)$ *be a natural chart for* TQ. *If* $c : I \to TU$ *is any differentiable curve (not necessarily self-consistent) and* $f : TQ \to R$ *is differentiable then*

$$(f\circ c)' = \sum_{j=1}^{k}\left(\frac{\partial f}{\partial\phi_j}\circ c\right)(\phi_j\circ\tau_Q\circ c)' + \sum_{j=1}^{k}\left(\frac{\partial f}{\partial\dot\phi_j}\circ c\right)(\dot\phi_j\circ c)'.$$

Proof. This follows as a special case of Theorem 7.2.10, on use of Definition 7.3.2. ■

The geometric interpretation of the next theorem is indicated in Figure 7.3.2. for the special case of a two-dimensional submanifold of R^3. The theorem has a traditional counterpart which is usually considered to be geometrically obvious.

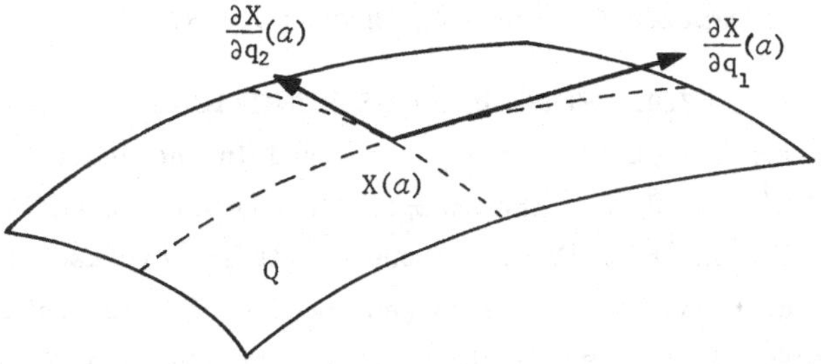

Figure 7.3.2.

7.3.10. Theorem. *If* Q *is a submanifold of* R^n *of dimension* k, X *the identity map on* R^n *and* (U, q) *a chart for* Q *at each* $a \in Q$, *then*

$$\left(a, \frac{\partial X}{\partial q_i}(a)\right) \in T_aQ$$

for $1 \leqslant i \leqslant k$.

Proof. Define $\gamma : I \to U$ by putting

$$\gamma(t) = q^{-1}(q(a) + t\, e_i)$$

with I an open interval chosen so that $0 \in I$ and $\gamma(I) \subseteq U$. Hence $[\gamma]_a \in T_aQ$. Since, moreover, γ maps I into R^n and q^{-1} maps an open set in R^k into R^n we get

$$\begin{aligned} \gamma'(0) &= (q^{-1})^{/i}\,(q(a)) \\ &= (X \circ q^{-1})^{/i} \circ q(a) \\ &= \frac{\partial X}{\partial q_i}(a). \quad \blacksquare \end{aligned}$$

EXERCISES 7.3.

1. *Prove Corollary 7.3.7.*
   You will need to use Theorem 7.3.6. and 7.3.3.
2. *Prove Corollary 7.3.8.*
3. *Note that the chart* $q = (\theta, r-\underline{1})$ *as defined in Example 6.1.4, is a chart for* R^2 *with the submanifold property for*

   $$Q = \{(a,b) : a^2 + b^2 = 1\}.$$

   Verify by direct calculation that Theorem 7.3.10. holds in this special case.
4. *Complete the proof of Theorem 7.3.9.*

7.4. HOMOGENEOUS FUNCTIONS

This topic is traditionally discussed in the context of functions which map R^n to R as, for example, in Courant & John(1974). For applications to mechanics, however, it is more useful to define homogeneity for functions having the tangent bundle of a manifold as domain. We suppose Q is a submanifold of R^n of dimension k.

7.4.1. Definition. A map $g: TQ \to R$ is said to be *homogeneous of degree* $r \geqslant 1$ if, for all $(a,v) \in TQ$ and $t \in R$

$$g(a, tv) = t^r g(a,v). \quad \blacksquare$$

For example, the map $g: TR^2 \to R$ with

$$g(x,y, h,k) = \cos(x)h^2 + 2hk + \sin(y)k^2$$

is homogeneous of degree 2.

The following theorem extends the familiar result known as "Euler's relation for homogeneous functions" to the new context.

7.4.2. Theorem. (Euler's relation). *If* $g: TQ \to R$ *is homogeneous of degree* r *and* $(TU, T\phi)$ *is a natural chart for* TQ *then*

$$\sum_{i=1}^{k} \dot{\phi}_i \frac{\partial g}{\partial \dot{\phi}_i} = r\, g.$$

Proof. Let $(a,h) \in TQ$ and define $f: R \to TQ$ by putting

$$f(t) = g(a,th).$$

Next define a curve $c: R \to TQ$ by putting

$$c(t) = (a,th).$$

Thus $f = g \circ c$ and so we may apply Theorem 7.3.9, to get

$$f' = \sum_{j=1}^{k} \left(\frac{\partial g}{\partial \phi_j} \circ c\right)(\phi_j \circ \tau_Q \circ c)' + \sum_{j=1}^{k} \left(\frac{\partial g}{\partial \dot{\phi}_j} \circ c\right)(\dot{\phi}_j \circ c)' .$$

But $\phi_j \circ \tau_Q \circ c$ is a constant function while $\dot{\phi}_j$ is linear on the tangent space T_aQ, hence $(\dot{\phi}_j \circ c)' = \dot{\phi}_j \circ (\underline{a},\underline{h})$. Thus

$$f' = \sum_{j=1}^{k} \left(\frac{\partial g}{\partial \dot{\phi}_j} \circ c\right)\dot{\phi}_j \circ (\underline{a},\underline{h}). \tag{1}$$

Now by homogeneity, $f(t) = t^r g(a,h)$ so that

$$f'(1) = rg(a,h) \tag{2}$$

Evaluating (1) at 1 and comparing with (2) shows that

$$\sum_{j=1}^{k} \frac{\partial g}{\partial \dot{\phi}_j} \dot{\phi}_j = r\, g$$

at each point (a,h) in the common domain of these two functions. $\blacksquare$

8. DERIVING LAGRANGE'S EQUATIONS

In this chapter we begin our account of classical mechanics. The physics background required is minimal. We will assume that the reader is familiar with the concepts of velocity, acceleration, mass and force at about the level of high school physics.

The results on partial derivatives established in the previous chapter will be used to derive Lagrange's equations for particle motion directly from Newton's laws. Our derivation follows the same general plan as the traditional one, which may be found in books such as Goldstein(1980), Synge & Griffith(1959) and Whittaker(1952). The use of manifolds, however, enables us to give a more geometrical slant to this topic in that we use ideas such as 'orthogonality with respect to a tangent space' in place of 'infinitesimal displacements' and 'virtual work'.

To give our calculations a conventional look we will use $X = (x,y,z)$ *for the identity chart on* $\mathbb{R}^3$ *and* $(X \circ \tau_Q, \dot{X})$ *for the naturally induced chart on* $T\mathbb{R}^3$.

8.1. LAGRANGE'S EQUATIONS FOR FREE-FALL

To provide some physical motivation we first derive Lagrange's equations for the very simple mechanical system consisting of a particle falling freely under gravity. Regard the force $\underset{\sim}{F}$ acting on a particle of mass m and the acceleration $\underset{\sim}{a}$ which it produces as vectors. Newton's laws then imply that

$$\underset{\sim}{F} = m\underset{\sim}{a}$$

provided measurements are made relative to an 'inertial frame of reference'. From the study of motion near the earth's surface it is often useful to suppose that axes fixed relative to the earth provide such a frame of reference; for the study of planetary motion, however, it would be more appropriate to fix the axes relative to the distant stars.

Consider now a particle of mass m moving near the surface of the earth and choose a cartesian coordinate system fixed relative to the earth with the x and y axes horizontal and the z axis vertical. We will assume that the particle is acted on by a constant force of magnitude mg vertically downwards. See Figure 8.1.1.

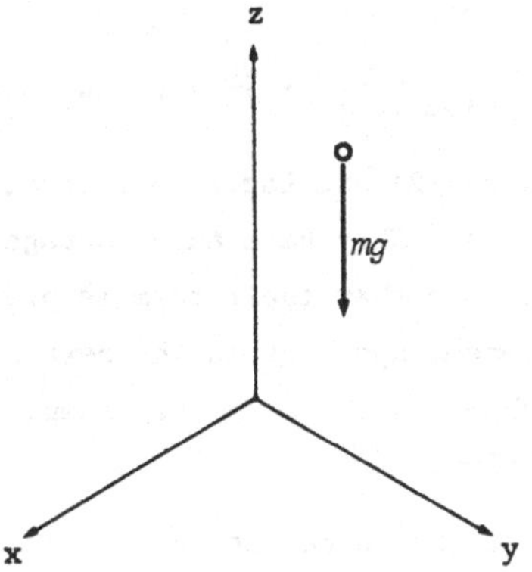

Figure 8.1.1. Motion under gravity.

It can be shown that the position and velocity of the particle at any instant determine its subsequent motion uniquely. Hence it is natural to represent its state at time t by an element $c(t) \in TR^3$ with $c: I \to TR^3$ a self-consistent path, which is defined on some interval I.

The cartesian coordinates of the particle at time t are $(X \circ \tau_{R^3} \circ c)(t)$ and so Newton's laws of motion imply that

$$m(x \circ \tau_{R^3} \circ c)''(t) = 0$$
$$m(y \circ \tau_{R^3} \circ c)''(t) = 0 \qquad (1)$$
$$m(z \circ \tau_{R^3} \circ c)''(t) = -mg.$$

Dynamically significant quantities associated with the motion of the particle are the *kinetic* and *potential energies*, given by the formulas

$$T = \tfrac{1}{2}m(\dot{x}^2 + \dot{y}^2 + \dot{z}^2)$$
$$V = mgz\ .$$

We regard T and V as maps from TR^3 and R^3, respectively, into R and then define the Lagrangian function $L : TR^3 \to R$ as

$$L = T - V \circ \tau_{R^3}\ .$$

where $\tau_{R^3} : TR^3 \to R^3$ is the natural projection. The equations (1) may be written in terms of L as

$$\begin{aligned}
\left(\frac{\partial L}{\partial \dot{x}} \circ c\right)' - \frac{\partial L}{\partial x} \circ c = 0 \\
\left(\frac{\partial L}{\partial \dot{y}} \circ c\right)' - \frac{\partial L}{\partial y} \circ c = 0 \qquad (2)\\
\left(\frac{\partial L}{\partial \dot{z}} \circ c\right)' - \frac{\partial L}{\partial z} \circ c = 0\ .
\end{aligned}$$

The equations (2) are Lagrange's equations for the motion of the free-falling particle. They have an advantage over the Newtonian equations of motion (1) in that their form is preserved under change of coordinates, as will become apparent in the next section. The gravitational force in the example above is the simplest example of a conservative field of force, as defined below.

8.1.1. Definition. If the force acting on a particle has the form

$$F(a) = \left(a, -\left(\frac{\partial V}{\partial x}, \frac{\partial V}{\partial y}, \frac{\partial V}{\partial z}\right)(a)\right) \in T_a R^3$$

at each $a \in U$, an open subset of R^3, for some differentiable function $V : U \to R$ we say that the map $F : U \to TR^3$ is *a conservative field of force*. ■

This definition has an obvious generalization to R^n. Further examples of conservative fields of force are given in the exercises.

EXERCISES 8.1.

1. *Derive Lagrange's equations (2) for the free-falling particle from their Newtonian form (1). You will first need to apply theorem 7.3.6 to the left hand sides of the equations (1).*

2. *Show that the gravitational field of force described in the text is conservative.*

3. *Show that the force field defined on* R^1 *by* $F(a) = (a, -ka)$ *is conservative. (This field of force arises, for* $k > 0$*, when Hooke's law is used to model the motion of a particle attached to an elastic spring and leads to the study of simple harmonic motion.)*

4. *Show that the force field defined on* $R^3\backslash\{0\}$ *by*

$$F(a,b,c) = \left((a,b,c),\ -\frac{1}{(\sqrt{a^2+b^2+c^2})^3}\ (a,b,c)\right)$$

is conservative. (This force field corresponds to the gravitatonal attraction produced by a particle of unit mass at the origin).

8.2. LAGRANGE'S EQUATIONS FOR A SINGLE PARTICLE

Lagrange's equations will now be derived for the motion of a particle of mass m which is constrained to move on a submanifold Q of R^3, the dimension of Q being k where $1 \leqslant k \leqslant 3$. Again we take $X = (x,y,z)$ to be the identity chart on R^3.

It will be assumed, furthermore, that the particle is subject to a conservative field of force together with a reaction force which acts orthogonally to the manifold at each point - as would be the case if there were no friction between the particle and the manifold. In mathematical terms this means that the force acting on the particle at each point $a \in Q$ is a sum

$$F(a) + R([\gamma]_a) = (F\circ\tau_Q + R)([\gamma]_a) \in TR^3$$

where $F(a)$ and $R([\gamma]_a)$ have the following forms:

$$F(a) = \left(a,\ -\left(\frac{\partial V}{\partial x}, \frac{\partial V}{\partial y}, \frac{\partial V}{\partial z}\right)(a)\right) \tag{1}$$

for some function $V : U \longrightarrow R$ where U is open, while the reaction force

$$R([\gamma]_a) \text{ is orthogonal to } T_aQ \tag{2}$$

in the sense of Definition 8.2.1. below.

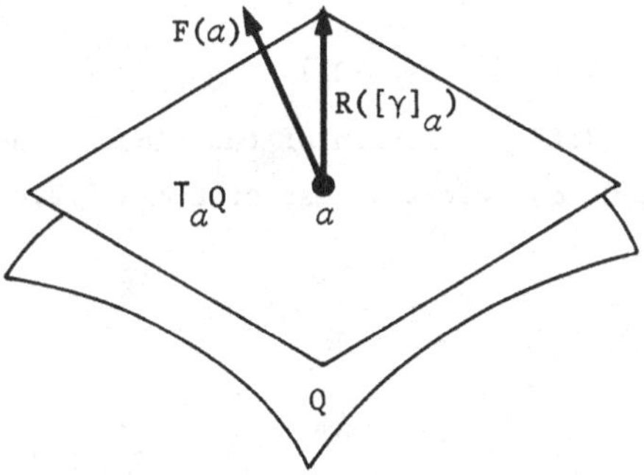

Figure 8.2.1.

8.2.1. Definition. A vector $(a,v) \in \mathsf{T}\mathbb{R}^n$ is *orthogonal* to the submanifold TQ of $\mathsf{T}\mathbb{R}^n$ at $a \in Q$ if the dot product $v.h = 0$ for every vector $(a,h) \in \mathsf{T}_a Q$. ■

At time t the position of the particle will be $\tau_Q \circ c(t)$ and its state will be $c(t)$ for some self-consistent path $c : I \to TQ$, where I is an interval. Relative to the cartesian coordinate system defined by the x,y and z axes the position and velocity of the particle will be $X \circ \tau_Q \circ c(t)$ and $\dot{X} \circ c(t)$ respectively. Hence Newton's laws imply that

$$
\begin{aligned}
m(X \circ \tau_Q \circ c)''(t) &= \pi_2 \circ (F \circ \tau_Q + R) \circ (X \circ \tau_Q, \dot{X}) \circ c(t) \\
&= \pi_2 \circ (F \circ \tau_Q + R) \circ c(t)
\end{aligned}
$$

where $\pi_2 : \mathbb{R}^3 \times \mathbb{R}^3 \to \mathbb{R}^3$ denotes projection onto the second factor. Hence by Theorem 7.3.6.

$$m(\dot{X} \circ c)'(t) = \pi_2 \circ (F \circ \tau_Q + R) \circ c(t). \tag{3}$$

To convert (3) to Lagrangian form we first introduce the *kinetic energy*

$$
\begin{aligned}
T &= \tfrac{1}{2} m(\dot{x}^2 + \dot{y}^2 + \dot{z}^2) \\
&= \tfrac{1}{2} m \dot{X}^2
\end{aligned}
$$

where $\dot{X}^2$ denotes $\dot{X} \cdot \dot{X}$. After restricting T to TQ as domain, we introduce the *Lagrangian function* $L : TQ \to \mathbb{R}$ by putting

$$L = T - V \circ \tau_Q \tag{4}$$

8.2.2. Theorem. (Lagrange's equations). *Under the above assumptions the curve* $c : I \to TQ$ *satisfies the equations*

$$\left(\frac{\partial L}{\partial \dot{q}_i} \circ c\right)' - \left(\frac{\partial L}{\partial q_i} \circ c\right) = 0 \qquad (i=1,\dots,k)$$

for each chart $(TU, ((q_1,\dots,q_k) \circ \tau_Q, (\dot{q}_1,\dots,\dot{q}_k)))$ *for* TQ.

Proof. Partial differentiation of the kinetic energy T with respect to each of q_i and $\dot{q}_i$ gives on use of Lemma 7.2.4, the semi-classical chain rule,

$$\frac{\partial T}{\partial q_i} = m\dot{X} \cdot \frac{\partial \dot{X}}{\partial q_i} \tag{5}$$

$$\frac{\partial T}{\partial \dot{q}_i} = m\dot{X} \cdot \frac{\partial \dot{X}}{\partial \dot{q}_i} = m\dot{X} \cdot \frac{\partial X}{\partial q_i} \circ \tau_Q \tag{6}$$

where the last equation follows from Corollary 7.3.4, the cancellation of dots rule.

Differentiation of (6) along the curve $c : I \longrightarrow TQ$ now gives

$$\left(\frac{\partial T}{\partial \dot{q}_i} \circ c\right)' = m\left[(\dot{X}\circ c)'\cdot\left(\frac{\partial X}{\partial q_i} \circ \tau_Q \circ c\right) + (\dot{X}\circ c)\cdot\left(\frac{\partial X}{\partial q_i} \circ \tau_Q \circ c\right)'\right]$$
$$= m\left[(\dot{X}\circ c)'\cdot\left(\frac{\partial X}{\partial q_i} \circ \tau_Q \circ c\right) + (\dot{X}\circ c)\cdot\left(\frac{\partial \dot{X}}{\partial q_i} \circ c\right)\right],$$

by Corollary 7.3.8, the interchange of dot and dash rule. Hence subtraction of (5) composed with c gives

$$\left(\frac{\partial T}{\partial \dot{q}_i} \circ c\right)' - \frac{\partial T}{\partial q_i} \circ c$$
$$= m(\dot{X} \circ c)' \cdot\left(\frac{\partial X}{\partial q_i} \circ \tau_Q \circ c\right)$$
$$= \pi_2\circ(F \circ\tau_Q + R)\circ c \cdot \left(\frac{\partial X}{\partial q_i} \circ \tau_Q \circ c\right) \text{ by Newton (3)}$$
$$= \pi_2\circ\left(F\cdot \frac{\partial X}{\partial q_i}\right)\circ \tau_Q \circ c \text{ by (2) and Theorem 7.3.10.}$$
$$= -\left(\frac{\partial V}{\partial X} \cdot \frac{\partial X}{\partial q_i}\right)\circ\tau_Q\circ c \qquad \text{by (1)}$$
$$= -\frac{\partial V}{\partial q_i} \circ \tau_Q \circ c \qquad \text{by Theorem 7.2.6.}$$

Thus $$\left(\frac{\partial T}{\partial \dot{q}_i} \circ c\right)' - \left(\frac{\partial T}{\partial q_i} \circ c\right) = -\frac{\partial V}{\partial q_i} \circ \tau_Q\circ c.$$

Lagrange's equations now follow from (4) since

$$\frac{\partial(V\circ\tau_Q)}{\partial q_i} = \frac{\partial V}{\partial q_i} \circ \tau_Q \text{ and } \frac{\partial(V\circ\tau_Q)}{\partial \dot{q}_i} = 0. \quad ■$$

8.3. LAGRANGE'S EQUATIONS FOR SEVERAL PARTICLES

Lagrange's equations will now be set up for a system of n particles each of which moves in R^3 subject to certain constraints. (One obvious constraint, for example, is that two particles may not simultaneously occupy the same position.) Here only those constraints will be considered which place restrictions upon the positions which the particle may occupy - and not those which place constraints upon their velocities, for example.

8.3.1. Definition. The *configuration set* of the system of particles will be defined as the set of all elements

$$a = (a_1, b_1, c_1, \ldots\ldots\ldots, a_n, b_n, c_n) \in \mathbb{R}^{3n}$$

such that for $i = 1,\ldots,n$ the ith particle may occupy the position $(a_i, b_i, c_i) \in \mathbb{R}^3$, in conformity with the constraints. ∎

In this way, instead of considering n particles moving in $\mathbb{R}^3$ we consider one particle moving in $\mathbb{R}^{3n}$. Only those systems will be considered in which the configuration set can be made into a submanifold Q of $\mathbb{R}^{3n}$.

8.3.2. Definition. If the configuration set of a system of n particles is a submanifold Q of $\mathbb{R}^{3n}$ then Q will be called the *configuration manifold* and the tangent bundle TQ will be called the *velocity phase space* of the system. ∎

Let the identity chart of $\mathbb{R}^{3n}$ be denoted by

$$X = (x_1, y_1, z_1, \ldots\ldots, x_n, y_n, z_n).$$

The history of the particles will then be described by a curve $c: I \to TQ$ such that the position of the ith particle at time t is $(x_i, y_i, z_i) \circ \tau_Q \circ c(t)$ and its velocity is $(\dot{x}_i, \dot{y}_i, \dot{z}_i) \circ c(t)$. We now generalize some of the ideas of Section 8.1.

8.3.3. Definition. The *field of force* for a system of n particles in $\mathbb{R}^3$ is the map $F : U \to \mathbb{R}^{3n}$ with

$$F(a) = (a,\ \pi_2 \circ F_1(a_1, b_1, c_1), \ldots\ldots, \pi_2 \circ F_n(a_n, b_n, c_n))$$

where $U \subseteq \mathbb{R}^{3n}$ is open and $F_i : U \to T\mathbb{R}^3$ is the field of force acting on the ith particle. The field of force F is said to be *conservative* if there exists a *potential function* $V : U \to \mathbb{R}$ such that

$$F(a) = \left(a,\ -\left(\frac{\partial V}{\partial x_1}, \frac{\partial V}{\partial y_1}, \frac{\partial V}{\partial z_1}, \ldots, \frac{\partial V}{\partial x_n}, \frac{\partial V}{\partial y_n}, \frac{\partial V}{\partial z_n}\right)(a)\right).\ ∎$$

8.3.4. Definition. Let $m_1,\ldots,m_n$ be the respective masses of the n particles of a system with configuration manifold $Q \subseteq \mathbb{R}^{3n}$. The *kinetic energy* of the system is the map $T : T\mathbb{R}^{3n} \to \mathbb{R}$ given by

$$T = \tfrac{1}{2} \sum_{i=1}^{n} m_i (\dot{x}_i^2 + \dot{y}_i^2 + \dot{z}_i^2).\ ∎$$

The above formula for the kinetic energy can be written more succinctly as a matrix product

$$T = \tfrac{1}{2}\dot{X}^T MX$$

where M is the diagonal matrix given in terms of the masses by

$$M = \text{diag}(m_1, m_1, m_1, m_2, m_2, m_2, \ldots\ldots, m_n, m_n, m_n)$$

and where $\dot{X}^T$ is the transpose of the column matrix $\dot{X}$.

We are now able to define the Lagrangian function of the system of particles.

8.3.5. Definition. Restricting the maps T and V in Definitions 8.3.4. and 8.3.3. to TQ we define the *Lagrangian function* $L : TQ \longrightarrow R$ by putting

$$L = T - V \circ \tau_Q. \quad ■$$

8.3.6. Theorem. *Consider a system of n particles in R^3 which are constrained to move in such a way that the equivalent single particle moves on a submanifold Q of R^{3n} of dimension k. Let the field of force acting on this particle be the sum of a conservative field of force and a reaction which is orthogonal to TQ at each point. Let $L : TQ \longrightarrow R$ be the Lagrangian function of the system. For each chart*

$$(TU, ((q_1, \ldots, q_k) \circ \tau_Q, (\dot{q}_1, \ldots, \dot{q}_k)))$$

for TQ, each self-consistent curve $c: I \longrightarrow TQ$ describing the history of the particle satisfies the equations

$$\left(\frac{\partial L}{\partial \dot{q}_i} \circ c\right)' - \frac{\partial L}{\partial q_i} \circ c = 0 \quad (1 \leqslant i \leqslant k).$$

Proof. The proof of this theorem is very similar to that of Theorem 8.2.2. Formally, one just replaces the real number " m " occurring in the proof of Theorem 8.2.2. by the diagonal matrix " M ". The validity of the resulting steps in the proof is established in the exercises. ■

EXERCISES 8.3.

The notation used in these exercises is explained above in the text.

1. *Check that the kinetic energy of the system, as given in Definition 8.3.4, can be written in the form claimed above,*

$$T = \tfrac{1}{2}\dot{X}^T M\dot{X}$$

2. *Check that Newton's law for the single particle equivalent to the system can be written in the form*

$$M(\dot{X}\circ c)'(t) = \pi_2\circ(F\circ\tau_Q + R)\circ c(t).$$

3. *Use the semi-classical chain rule, Lemma 7.2.4, to show that for* $1 \leq i \leq k$,

(a) $$\frac{\partial T}{\partial q_i} = \dot{X}^T M \frac{\partial \dot{X}}{\partial q_i}$$

(b) $$\frac{\partial T}{\partial \dot{q}_i} = \dot{X}^T M \frac{\partial X}{\partial q_i} \circ \tau_Q.$$

4. *Show that* $(T\circ c)' = (\dot{X}\circ c)'^T M(\dot{X}\circ c)$

5. *Use the results of Exercise 2 and 3 to give a proof of Theorem 8.3.6.*

8.4. MOTION OF A RIGID ROD

In the previous section we derived Lagrange's equations for a conservative system for which the resulting single particle was constrained to move on a submanifold Q of R^{3n}. We now show how it is possible to fit a prototype rigid body into this scheme.

It will consist of two particles constrained to move in such a way that the distance between them remains constant. The reaction forces between the two particles are assumed to be equal in magnitude but opposite in direction and to act along the line joining the two particles, as in Figure 8.4.1. This system is thus a mathematical model of a pair of particles joined by a rigid rod of negligible mass.

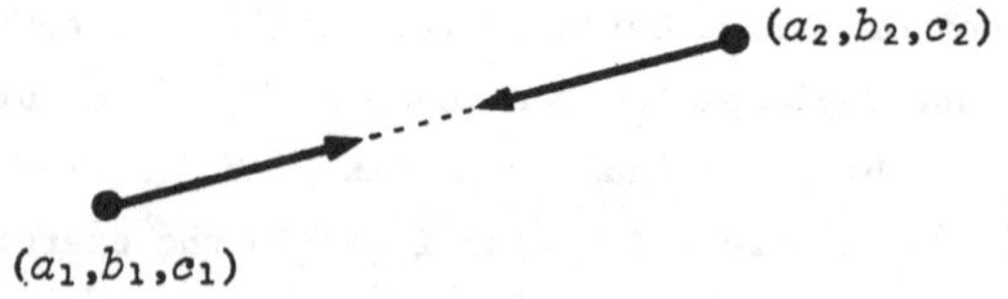

Figure 8.4.1. A rigid rod.

The *configuration set* for this system is a set of points of the form

$$a = (a_1, b_1, c_1, a_2, b_2, c_2) \in R^6$$

where (a_1,b_1,c_1) and (a_2,b_2,c_2) are the positions of the two particles in R^3. The constraint that the two particles lie at opposite ends of a rigid rod of length $d > 0$, can then be expressed by the condition that

$$f(a) = 0$$

where $f: R^6 \longrightarrow R$ with

$$f(a) = (a_1-a_2)^2 + (b_1-b_2)^2 + (c_1-c_2)^2 - d^2 \tag{1}$$

We now show that the configuration set is a submanifold of R^6. Let $a \in R^6$ be as above and let $h \in R^6$. By (1) the Fréchet derivative of f is given by the dot product

$$Df(a)(h) = 2(a_1-a_2, b_1-b_2, c_1-c_2, a_2-a_1, b_2-b_1, c_2-c_1) \cdot h \tag{2}$$

Clearly $Df(a)$ has full rank 1 and so Theorem 3.3.1, shows that *the configuration set* $f^{-1}(0)$ *is a submanifold of* R^6, *of dimension* 5.

The following proposition states a somewhat surprising fact, which is crucial for the valid application of Theorem 8.3.6.

8.4.6. Proposition. *The reaction force due to the rigid rod connecting two particles is orthogonal to the configuration manifold at each point.*

Proof. Let Q denote the configuration manifold $f^{-1}(0)$, with f given by (1) above, and let $(a,h) \in T_aQ$. By Theorem 5.3.1,

$$Df(a)(h) = 0$$

which by (2), is equivalent to

$$2(a_1-a_2, b_1-b_2, c_1-c_2, a_2-a_1, b_2-b_1, c_2-c_1)' \cdot h = 0. \tag{3}$$

Since the reaction forces on the two particles act along the rigid rod and are equal in magnitude but opposite in direction, we may write their directions as

$$\lambda(a_1-a_2, b_1-b_2, c_1-c_2) \quad \text{and} \quad -\lambda(a_1-a_2, b_1-b_2, c_1-c_2)$$

respectively. These two reaction forces are combined to give a single reaction force in T_aR^6

$$R = (a, \lambda(a_1-a_2, b_1-b_2, c_1-c_2, a_2-a_1, b_2-b_1, c_2-c_1))$$

and so, by (3), $R.(a,h) = 0$. Thus the reaction R at the point a is orthogonal to T_aQ.

If now we are given a potential function $V : R^6 \longrightarrow R$ for the pair of particles and a submanifold chart $(U,(q_1,q_2,q_3,q_4,q_5))$, for Q we may apply Theorem 8.3.6, to obtain Lagrange's equations for the system.

EXERCISE 8.4.

Find a submanifold chart suitable for the configuration manifold (Hint: fix one end of the rod and then find a submanifold chart for $S^2 \subset R^3$.)

8.5. CONSERVATION OF ENERGY

Physical systems are subject to the law of conservation of energy. For systems of the type discussed in Section 8.3, it will be shown from Newton's laws that the total mechanical energy - kinetic energy plus potential energy - remains constant with the lapse of time. From a physical viewpoint, these systems conserve their mechanical energy because they are free from the effects of such forces as friction and air-resistance, which dissipate mechanical energy into other forms of energy such as heat.

Returning to the situation described in Section 8.3, we consider a single particle moving on a submanifold Q of R^{3n} subject to a conservative field of force together with a reaction force orthogonal to TQ at each point. As before, we let $V : Q \longrightarrow R$ be a potential function for, and $T : TQ \longrightarrow R$ be the kinetic energy of, the particle. The *total energy function* $E : TQ \longrightarrow R$ for the system is defined by putting

$$E = T + V \circ \tau_Q .$$

On the basis of Newton's laws, the following theorem can now be proved.

8.5.1. Theorem. (Conservation of energy). *If* $c: I \longrightarrow TQ$ *is a self-consistent curve describing the history of the above particle then* $E \circ c$ *is a constant function.*

Proof. The notation being as in Section 8.3,

$$(V \circ \tau_Q \circ c)' = \left(\frac{\partial V}{\partial X} \circ \tau_Q \circ c\right) \cdot (\dot{X} \circ c) \qquad \text{by Corollary 7.3.7.}$$

$$= -\left(\pi_2 \circ F \circ \tau_Q \circ c\right) \cdot (\dot{X} \circ c) \qquad \text{by Definition 8.3.3.}$$

$$= -(\pi_2 \circ (F \circ \tau_Q + R) \circ c) \cdot (\dot{X} \circ c)$$

where the last step is valid since, by Theorem 7.3.6, $(\dot{X} \circ c)$ equals $(X \circ \tau_Q \circ c)'$, which lies in TQ and hence is orthogonal to the reaction R at each point. Note also that

$$(T\circ c)' = (\dot{X}\circ c)'^{T}M\,(\dot{X}\circ c) \qquad \text{by Exercise } 8.3.4.$$

$$= (\pi_2\circ(F\circ\tau_Q + R)\circ c)\cdot(\dot{X}\circ c) \quad \text{by Exercise } 8.3.2.$$

Thus $(E\circ c)' = \underline{0}$ in the interval I and $E\circ c$ is a constant. ■

The above proof is a rigorous version of one of the two proofs given in Whittaker(1952) section 41. In the alternative approach Whittaker first of all proves a conservation law for a system of Lagrangian equations defined in terms of an arbitrary Lagrangian function L and then specializes the result to the case where L may be expressed in terms of kinetic and potential energies. In the exercises, the reader is invited to follow through the details of the alternative approach by way of comparison.

EXERCISES 8.5.

In these exercises (U, q) *denotes a chart for a k-dimensional submanifold* Q *of* R^{3n}.

1. *Let* $L : TQ \to R$ *be any differentiable function and suppose that* $c : I \to TU$ *is a self-consistent path satisfying Lagrange's equations*

$$\left(\frac{\partial L}{\partial \dot{q}_i}\circ c\right)' - \frac{\partial L}{\partial q_i}\circ c = 0 \quad \textit{for } 1 \leqslant i \leqslant k.$$

Show, by using Theorems 7.3.9. and 7.3.6, that the derivative of the function

$$\left(\sum_{i=1}^{k} \dot{q}_i \frac{\partial L}{\partial \dot{q}_i} - L\right)\circ c$$

is zero on I *and hence that this function is a constant.*

2. *Let* $T: TQ \to R$ and $V : Q \to R$ *be differentiable functions and suppose that* T *is homogeneous of degree* 2 *and let*

$$L = T - V\circ\tau_Q .$$

(a) *Show, from Theorem 7.4.2, that*

$$\sum_{i=1}^{k} \dot{q}_i \frac{\partial L}{\partial \dot{q}_i} - L = T + V\circ\tau_Q$$

(b) *Hence prove Theorem 8.5.1.*

9. FORM OF LAGRANGE'S EQUATIONS

This chapter begins with a simple example of a mechanical system. By explicit calculations, the Lagrange equations for this system can be proved equivalent to a system of second-order differential equations.

For arbitrary mechanical systems of the type studied in Chapter 8, a similar fact can be proved by using standard results about quadratic forms. In subsequent chapters the theory of differential equations will be used to show how Lagrange's equations determine the motion of mechanical systems.

9.1. MOTION ON A PARABOLOID

This section investigates the explicit form of Lagrange's equations for a mechanical system consisting of a particle of mass m moving under gravity on a paraboloid in the absence of friction. The paraboloid is assumed to be the subset Q of R^3 consisting of all the points on which the equation

$$z = \tfrac{1}{2}(x^2 + y^2) \qquad (1)$$

holds, where (x, y, z) is the identity map on R^3. Thus Q is a paraboloid of circular cross-section with its axis vertical, as illustrated in Figure 9.1.1.

A chart for R^3 with the submanifold property for Q is

$$(R^3, \ (x, y, z - \tfrac{1}{2}(x^2 + y^2))).$$

Hence Q is a submanifold of R^3 and has (Q, (x,y)) as a chart. Incidentally, this chart forms a one-chart atlas for Q and so Lagrange's equations with respect to this chart apply to the whole of the tangent bundle TQ.

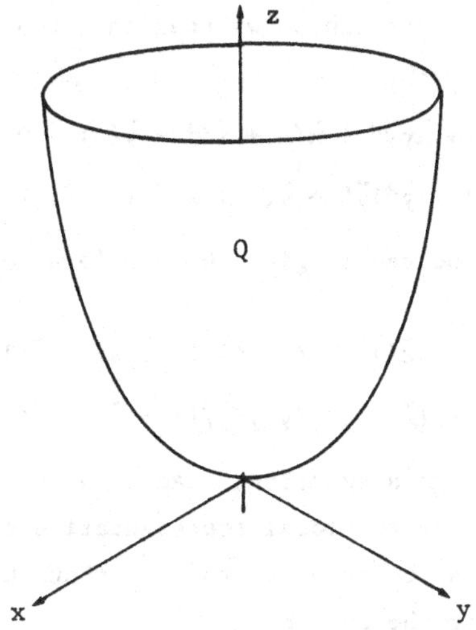

Figure 9.1.1. A paraboloid Q.

With x,y and z restricted to Q it follows from (1) and Theorem 7.3.3. that

$$\dot{z} = (x\circ\tau_Q)\dot{x} + (y\circ\tau_Q)\dot{y} .$$

Hence, on temporarily writing x,y in place of $x\circ\tau_Q$ and $y\circ\tau_Q$ for simplicity, we find that the kinetic energy is given by

$$T = \tfrac{1}{2}m(\dot{x}^2 + \dot{y}^2 + (x\dot{x} + y\dot{y})^2).$$

Since the potential function, moreover, is simply $V = mg\,z$ it follows that the Lagrangian function is given by

$$L = \tfrac{1}{2}m\ ((1+x^2)\dot{x}^2 + 2xy\ \dot{x}\dot{y} + (1+y^2)\dot{y}^2) - \tfrac{1}{2}mg(x^2+y^2). \tag{2}$$

Now let $c: I \to TQ$ be a self-consistent curve describing the history of the particle. We assume c is of class $\mathbb{C}^2$ and introduce the notation

$$\bar{x} = x\circ\tau_Q\circ c, \qquad \bar{y} = y\circ\tau_Q\circ c \tag{3}$$

for the local representative $(\tau_Q\circ c)_{(x,y)}$ of the curve $\tau_Q\circ c$ with respect to the chart (Q, (x,y)). From the interchange of dot and dash rule, Theorem 7.3.6, it follows that

$$\bar{x}' = \dot{x}\circ c, \quad \bar{y}' = \dot{y}\circ c. \tag{4}$$

An application of Theorem 8.2.2. shows that Lagrange's equations are satisfied by the curve c and from (2), (3) and (4) it follows that these equations can be written in terms of the local representatives as

$$\begin{aligned} (1 + \bar{x}^2)\bar{x}'' + \bar{x}\bar{y}\bar{y}'' + \bar{x}(g + \bar{x}'^2 + \bar{y}'^2) &= 0 \\ \bar{x}\bar{y}\,\bar{x}'' + (1 + \bar{y}^2)\bar{y}'' + \bar{y}(g + \bar{x}'^2 + \bar{y}'^2) &= 0 \end{aligned} \tag{5}$$

These equations can be solved to give the acceleration $\bar{x}''$ and $\bar{y}''$ explicitly:

$$\begin{aligned} \bar{x}'' &= -\bar{x}[g + (\bar{x}')^2 + (\bar{y}')^2]/[1 + \bar{x}^2 + \bar{y}^2] \\ \bar{y}'' &= -\bar{y}[g + (\bar{x}')^2 + (\bar{y}')^2]/[1 + \bar{x}^2 + \bar{y}^2]. \end{aligned} \tag{6}$$

Thus Lagrange's equations lead to a pair of second-order differential equations for the local representative of the curve c. Conversely, the steps can be reversed to show that these equations imply Lagrange's equations for the system.

EXERCISES 9.1.

1. *For the problem discussed in the text concerning motion on the paraboloid* Q:

(a) *Show that the kinetic energy is given by*

$$T + \tfrac{1}{2}m(\dot{x}\ \dot{y})A\circ\tau_Q \begin{pmatrix}\dot{x}\\ \dot{y}\end{pmatrix}$$

for a suitable matrix-valued function $A : Q \longrightarrow R^{2\times 2}$ *with pointwise inverse* $A^{-1} : Q \longrightarrow R^{2\times 2}$.

(b) *Show that Lagrange's equations (5) for the paraboloid can be written as:*

$$(A\circ\tau_Q\circ c)\begin{pmatrix}\bar{x}''\\ \bar{y}''\end{pmatrix} + (g + \bar{x}'^2 + \bar{y}'^2)\begin{pmatrix}\bar{x}\\ \bar{y}\end{pmatrix} = \begin{pmatrix}0\\0\end{pmatrix}$$

and then deduce the differential equations (6) with the aid of A^{-1}.

2. *Show that the system of second-order differential equations (6), describing motion on the paraboloid, admits the family of periodic solutions given by*

$$\bar{x}(t) = a\cos(\sqrt{g}\,t),\quad \bar{y}(t) = a\sin(\sqrt{g}\,t)$$

for each $a \in R$. *Note that the period is constant along the family. Use the relations (3) in the text to help you sketch the set of points*

$$\{\tau_Q \circ c(t) : t \in R\} \subseteq Q$$

which is traced out by the particle on Q.

3. *(The spherical pendulum). Consider a particle of mass* m *moving under gravity on the unit sphere* $S^2 \subseteq R^3$ *in the absence of friction. Let* $(U, (\theta,\phi))$ *be the spherical-polar chart for* S^2 *defined in Exercise 2.1.6.*

(a) *Show that the kinetic and potential energies may be expressed in this chart as*

$$T = \tfrac{1}{2}m(\dot{\theta}^2 + \sin^2\circ\theta\ \dot{\phi}^2)$$

$$V = mg\cos\circ\theta.$$

(b) *Verify that, with* $\bar{\theta} = \theta\circ\tau_Q\circ c$ *and* $\bar{\phi} = \phi\circ\tau_Q\circ c$, *Lagrange's equations are explicity*

$$m\left(\bar{\theta}'' - (\sin\circ\bar{\theta})(\cos\circ\bar{\theta})\bar{\phi}'^2 - g\sin\circ\bar{\theta}\right) = 0$$

$$(m(\sin^2\circ\bar{\theta})\bar{\phi}')' = 0.$$

(c) *Deduce that Lagrange's equations are equivalent to a system of second-order differential equations in* $\bar{\theta}$ *and* $\bar{\phi}$.

(d) *Show that while the particle stays within the chart domain* U, $(\sin^2\circ\bar{\theta})\bar{\phi}'$ *remains constant.*

9.2. QUADRATIC FORMS

The discussion of Lagrange's equations in the next section will use the basic ideas about quadratic forms which are summarized below. A fuller account of this topic is contained in standard linear algebra texts such as Nering(1964). Our definition of the matrix of a quadratic form differs from the usual one, however, and is more natural in the context of manifolds.

A *quadratic form* on a vector space V over R is a mapping $g: V \to R$ whose values are given by

$$g(v) = f(v,v)$$

for some bilinear map $f: V\times V \to R$. If g is a quadratic form then there is only one such bilinear form which is symmetric and we call it *the bilinear form generating* g.

A quadratic form g is said to be *positive definite* if $g(v) > 0$ for all $v \neq 0$. It is clear that the restriction of a positive definite quadratic form to a vector subspace of its domain is also a positive definite quadratic form.

9.2.1. Example. We claim that the map $g: R^2 \to R$ with values given by

$$g(a,b) = a^2 + 6ab + 10b^2$$

is a quadratic form and is positive definite. To see this it is helpful to use matrices and write

$$g(a,b) = (a\ b) \begin{pmatrix} 1 & 3 \\ 3 & 10 \end{pmatrix} \begin{pmatrix} a \\ b \end{pmatrix}.$$

We construct the bilinear map f generating g by putting

$$f((a,b),\ (c,d)) = (a\ b) \begin{pmatrix} 1 & 3 \\ 3 & 10 \end{pmatrix} \begin{pmatrix} c \\ d \end{pmatrix}.$$

Hence g is a quadratic form. Completing the square shows that

$$g(a,b) = (a + 3b)^2 + b^2 \geqslant 0$$

and this vanishes only if $(a,b) = (0,0)$. Hence g is positive definite. ■

The representation of quadratic forms by matrices will play an important rôle in the next section. The relevant definitions and theorems will be stated first for the special case of quadratic forms on R^n. The usual basis for R^n will be denoted by $(e_1,\ldots,e_n)$.

9.2.2. Definition. The *matrix of a quadratic form* g *on* R^n has as its i-jth element

$$f(e_i,e_j)$$

where f is the bilinear function generating g. ■

9.2.3. Theorem. *If* g *is a quadratic form on* R^n *with matrix* A *then for each* $v \in R^n$, *regarded as a column matrix,*

$$g(v) = v^T Av.$$

Conversely if this relation holds for all v *then* g *is a quadratic form and* A, *if symmetric, is the matrix of* g. ■

More generally, we now consider quadratic forms on an arbitrary vector space V of dimension n over R. There is then a vector-space isomorphism $\phi: V \to R^n$, which we may regard as a chart for V.

9.2.4. Theorem. *If* $g\colon V \to R$ *is a quadratic form on* V *then its local representative* g_ϕ *(shown in Figure 9.2.1.) is a quadratic form on* R^n. *If* A *denotes the matrix of* g_ϕ *then the values of* g *are given by*

$$g(v) = \phi(v)^T A\phi(v)$$

where we regard $\phi(v)$ *as a column matrix. Conversely if the values of* g *are given by this formula then* g *is a quadratic form on* V. ∎

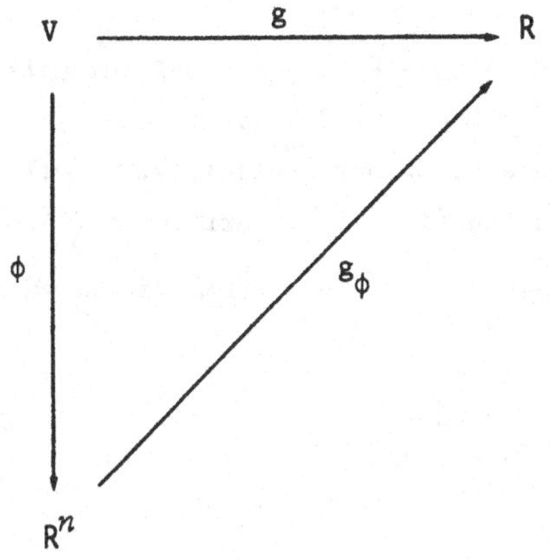

Figure 9.2.1. Local representative of a quadratic form.

The key theorem for our purposes is as follows:

9.2.5. Theorem. *If a quadratic form* $g\colon V \to R$ *is positive definite then the matrix of each local representative is non-singular.* ∎

9.3. LAGRANGE'S EQUATIONS ARE SECOND-ORDER.

It will now be shown how Lagrange's equations for a system of the type discussed in Section 8.3 can be written as a system of second-order differential equations. The kinetic energy of the system of particles was there defined in terms of the identity chart $(X\circ\tau, \dot{X})$ on TR^{3n} as

$$T \doteq \tfrac{1}{2}\, \dot{X}^T M\, \dot{X}$$

where $M = \mathrm{diag}(m_1, m_1, m_1, m_2, m_2, m_2, \ldots, m_n, m_n, m_n)$ with each mass $m_i > 0$. Thus for each $a \in R^{3n}$ the restriction of T to T_aR^{3n} is a positive-definite quadratic form.

Now suppose that the particle representing the system is constrained to move on a smooth k-dimensional submanifold Q of R^{3n}. For each $a \in Q$ the restriction of T to T_aQ is the restriction of a positive-definite quadratic form to a vector subspace of its domain, hence it is again a positive-definite quadratic form. Suppose furthermore, that $(q\circ\tau_Q, \dot{q})$ is a chart for TQ so that $\dot{q}|T_aQ$ is a linear chart on T_aQ. It then follows from Theorem 9.2.4. that the kinetic energy for the particle moving on Q is given by

$$T = \tfrac{1}{2}\dot{q}^T A\circ\tau_Q\,\dot{q} \tag{1}$$

for some matrix-valued function $A:Q \to R^{k\times k}$. A simple example illustrating this is contained in Exercise 9.1.1.

To work out Lagrange's equations explicitly, we will need the smoothness of the matrix valued function A occurring in (1).

9.3.1. Lemma. *The i-jth entry* a_{ij} *of the matrix-valued function* A *in* (1) *above is given by*

$$a_{ij}\circ\tau_Q = \frac{\partial^2 T}{\partial\dot{q}_i\,\partial\dot{q}_j}$$

for each $i,j = 1,2,\ldots,k$.

Proof. Notice that equation (1), written out in full, is

$$T = \tfrac{1}{2}\sum_{\ell=1}^{k}\sum_{m=1}^{k} a_{\ell m}\circ\tau_Q\,\dot{q}_\ell\,\dot{q}_m\,.$$

Differentiating this expression gives the required result. ■

9.3.2. Corollary. *The matrix-valued function* A *is smooth and*

$$\left(\frac{\partial T}{\partial\dot{q}}\circ c\right)' = A\circ\tau_Q\circ c\,(q\circ\tau_Q\circ c)'' + (A\circ\tau_Q\circ c)'(q\circ\tau_Q\circ c)'$$

for each smooth self-consistent path $c: I \to TQ$. ■

We can now prove the main result of this chapter.

9.3.3. Theorem. *Lagrange's equations are a system of second-order differential equations in the local representative* $\bar{q} = q\circ\tau_Q\circ c$.

Proof. We may suppose the Lagrangian function $L : TQ \to R$ is given by

$$L = \tfrac{1}{2}\dot{q}^T A\circ\tau_Q\,\dot{q} - V\circ\tau_Q$$

where A is as in Lemma 9.3.1. By Theorem 8.3.6. Lagrange's equations in a chart $(q\circ\tau_Q, \dot{q})$ become

$$\left(\frac{\partial}{\partial \dot{q}}\,(\tfrac{1}{2}\dot{q}^T A\circ\tau_Q\,\dot{q})\circ c\right)' - \frac{\partial L}{\partial q}\circ c = 0$$

where $c: I \longrightarrow TQ$ is a smooth self-consistent path.

Application of Corollary 9.3.2. then gives

$$(A\circ\tau_Q\circ c)\;(q\circ\tau_Q\circ c)'' + (A\circ\tau_Q\circ c)'\;(q\circ\tau_Q\circ c)' - \frac{\partial L}{\partial q}\circ c = 0.$$

By Theorem 9.2.5. the matrix-valued function $A\circ\tau_Q\circ c$ is pointwise invertible so the above equation becomes

$$(q\circ\tau_Q\circ c)'' = (A\circ\tau_Q\circ c)^{-1}\;(-(A\circ\tau_Q\circ c)'(q\circ\tau_Q\circ c)' + \frac{\partial L}{\partial q}\circ c)$$

as required. ■

EXERCISES 9.3.

1. *Fill in the details of the proof of Lemma 9.3.1.*

2. *Compare the contents of Section 9.3 with Sections 27 and 28 of Whittaker(1952).*

10. VECTORFIELDS

In modern terminology, the right-hand side of a differential equation determines a "vectorfield", which means assigning an "arrow" to each point of the domain, while the solutions of the differential equation are referred to as "integral curves", which are tangential to the arrows. In the context of manifolds, the integral curves are regarded as maps into the manifold while the "arrows" are elements of its tangent spaces. The modern terminology thus invites us to think of differential equations in a very geometrical way. In this chapter, enough of the theory of vectorfields will be developed to enable Lagrange's and Newton's equations to be put into proper geometric and analytic perspective.

Among the most basic results in the theory of differential equations are those concerning the existence and uniqueness of solutions and these results translate easily into corresponding results about integral curves. When studying vectorfields and their integral curves on a manifold, however, it is important to distinguish between "local" and "global" properties; the former refer to what happens in a particular chart domain whereas the latter refer to what happens on the whole manifold.

Finally, some special properties of vectorfields which arise from second-order differential equations, such as Lagrange's equations, will be studied.

10.1. BASIC IDEAS

In this section we introduce the ideas of vectorfields and integral curves and give a vectorfield version of the classical existence and uniqueness theorem for the solutions of differential equations.

A vectorfield may be thought of as a mapping from a manifold M to its tangent space TM which preserves the base points. The formal definition is as follows.

10.1.1. Definition. Let M be a smooth manifold. A mapping

$$Y : M \longrightarrow TM$$

with the property $\tau_M \circ Y = id_M$ is called *a vectorfield on* M. ■

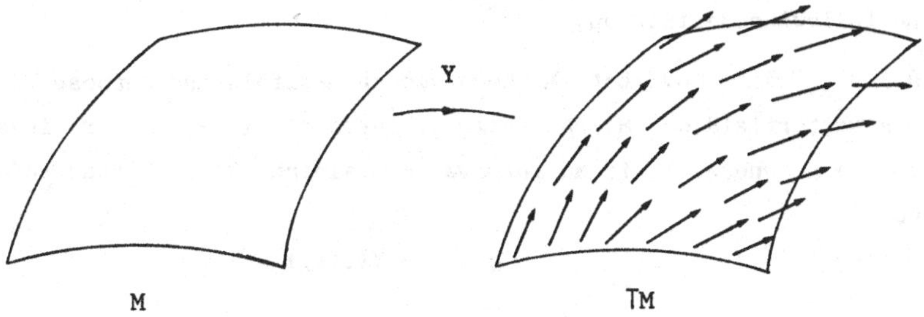

Figure 10.1.1. A vectorfield on M.

10.1.2. Example. The mapping

$$Y : R^2 \longrightarrow TR^2 \text{ given by}$$

$$Y(x,y) = ((x,y), (-y,x))$$

is a vectorfield on R^2 since $\tau_{R^2} \circ Y = id_{R^2}$. The assignment of the various arrows to points in R^2 is illustrated in Figure 10.1.2. ■

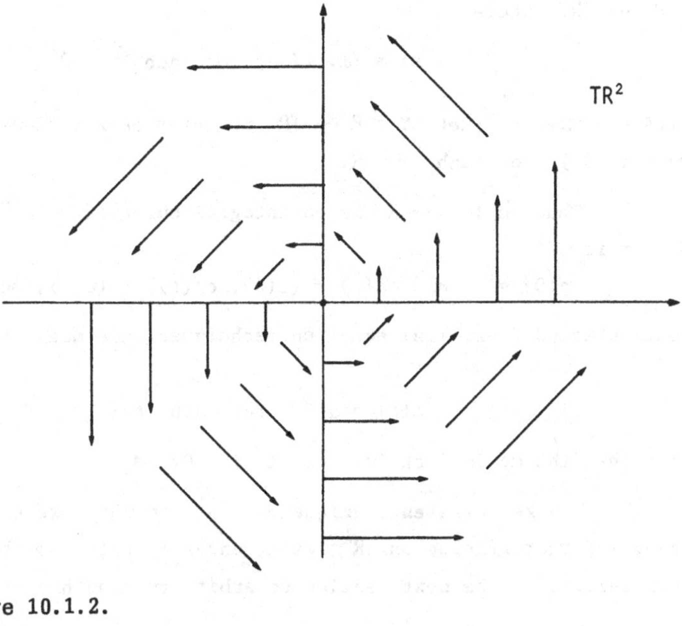

Figure 10.1.2.

Before reading on, recall (see Definition 7.1.1.) that if $c: I \to M$ is differentiable (I an interval, M a manifold), then we define

$$c^{\cdot}(t) = Tc(t,1) = (c(t), c'(t)).$$

The manifold version of a solution to a differential equation is given by the following definition.

10.1.3. Definition. Let M be a smooth manifold and suppose $Y : M \to TM$ is a vectorfield on M. An *integral curve of* Y *at* $a \in M$ is a mapping $c: I \to M$ where I is an interval containing 0 such that $c(0) = a$ and

$$c^{\cdot}(t) = Y(c(t))$$

for each $t \in I$. ■

The relationship between a differential equation on R^k written as

$$c'(t) = f(c(t)) \tag{1}$$

for some function $f : R^k \to R^k$ and its vectorfield counterpart is then given by including the "base point" as follows:

$$c^{\cdot}(t) = (c(t), c'(t)) = (c(t), f(c(t))).$$

Thus the vectorfield for the differential equation (1) is then given by $Y : R^k \to TR^k$ where

$$Y(x) = (x, f(x)) \text{ for each } x \in R^k.$$

10.1.4. Example. Let $Y : R \to TR$ be the vectorfield defined by $Y(x) = (x, 3x)$ for each $x \in R$.

Thus $c : I \to R$ is an integral curve at $a \in R$ for Y if and only if

$$c(0) = a \text{ and } c^{\cdot}(t) = (c(t), c'(t)) = (c(t), 3c(t)).$$

By elementary differential equation techniques this means that $c: I \to R$ with

$$c(t) = ae^{3t} \text{ for each } t \in I.$$

Notice that the maximal choice for I is R. ■

The key existence-uniqueness theorem which we state below in the language of vectorfields on R^k will, under certain smoothness assumptions, be generalized in the next section to arbitrary smooth manifolds.

10.1.5. Theorem. *Let* U *be an open subset of* R^k *and let* $F : U \to TU$ *be a vectorfield on* U. *Suppose that there exists a* $K > 0$ *such that for each* $y,z \in U$

$$\|\pi_2 \circ F(y) - \pi_2 \circ F(z)\| \leqslant K \|y - z\|. \tag{2}$$

Then for each $a \in U$, *there is an* $\varepsilon > 0$ *and a unique* $c: (-\varepsilon,\varepsilon) \to U$ *such that for each* $t \in (-\varepsilon,\varepsilon)$

$$c(0) = a \quad \text{and} \quad c^{\boldsymbol{\cdot}}(t) = F(c(t)).$$

Proof. See Chillingworth(1976) or any standard analysis text. ■

10.1.6. Lemma. *The condition (2) in Theorem 10.1.5. can be replaced by "*F *is* C^1 *on* U*".* ■

EXERCISES 10.1.

1. *Show that the vectorfield* $Y : R \to TR$ *given by* $Y(x) = (x,x^2)$ *has a unique integral curve at* 1 *which is defined on a maximal open interval. Show nevertheless that the solution is not defined on the whole of* R.

2. *Show that the vectorfield* $Y : R^+ \to TR^+$ *given by* $Y(x) = (x, -x^{\frac{1}{2}})$ *has as integral curves at* 0 *the zero function* $\underline{0}$ *and the function* $c: R \to R^+$ *given by*
$$c(t) = \begin{cases} (1+\tfrac{1}{2}t)^2, & t < -2 \\ 0, & t \geqslant -2. \end{cases}$$
Explain why this does not contradict Theorem 10.1.5.

3. *Find the integral curve for the vectorfield of Example 10.1.2. at the point* (a,b) *of* R^2. *What is the largest open interval in* R *on which it may be defined?*

10.2. MAXIMAL INTEGRAL CURVES

In Section 10.1. we showed that a smooth vectorfield on an open subset of R^k gave rise to integral curves which were defined on some interval I. In this section we show how that theory may be applied to obtain integral curves on the manifold itself. Finally we show how integral curves may be patched together in an unambiguous way. The idea is as follows:

take local representatives of the vectorfield on the manifold
↓
obtain integral curves in R^k *of the resulting vectorfields on* R^k
↓
lift these integral curves up to the manifold via the inverses of the chart mappings
↓
show that these integral curves may be patched together to give unique maximal integral curves on the manifold.

We begin this process with the manifold version of the existence-uniqueness Theorem 10.1.5. The relationship between a vectorfield and its local representative is illustrated in Figure 10.2.1.

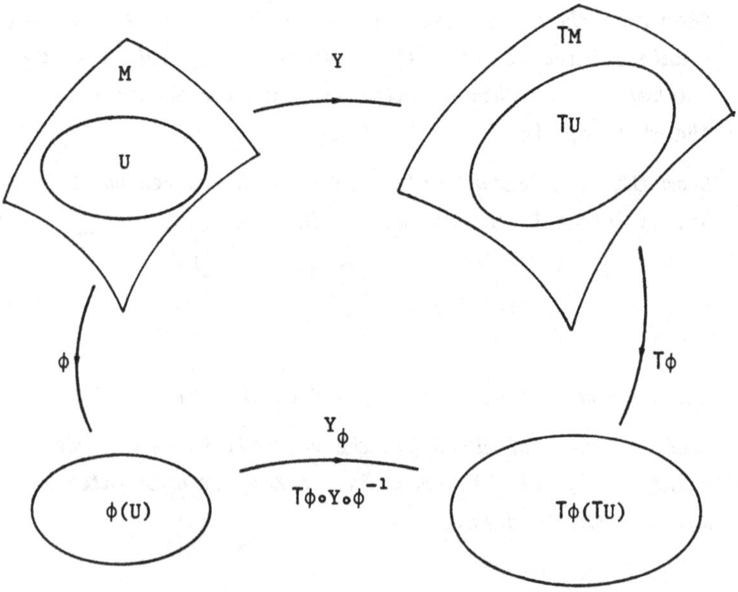

Figure 10.2.1. Local representative Y_ϕ of a vectorfield Y

10.2.1. Theorem. *Let* M *be a smooth k-dimensional submanifold of* R^n *and suppose* $Y : M \to TM$ *is a smooth vectorfield on* M.

Then for each $a \in M$ *there is an* $\varepsilon > 0$ *and a unique integral curve* $c : (-\varepsilon,\varepsilon) \to M$ *of* Y *at* a.

Proof. Suppose (U,ϕ) is a chart for M with $a \in U$. Then by Theorem 10.1.5. and Lemma 10.1.6. there is an integral curve $\xi : (-\varepsilon,\varepsilon) \to \phi(U)$ such that

$$\dot{\xi}(t) = T\phi \circ Y \circ \phi^{-1}(\xi(t))$$

with

$$\xi(0) = \phi(a).$$

Setting $c = \phi^{-1} \circ \xi$ gives the existence of a suitable integral curve on M.

For uniqueness see the exercises. ∎

We now show how the domain of each integral curve may be extended to yield "maximal" integral curves.

10.2.2. Lemma. *Let* $Y : M \to TM$ *be a smooth vectorfield on a smooth* k*-dimensional submanifold* M *of* R^n *and suppose*

$$\{c_\lambda : I_\lambda \to M \mid \lambda \in J\}$$

is the set of all integral curves of Y *at* $a \in M$. *Then for each pair* $\lambda,\mu \in J$ *and each* $t \in I_\lambda \cap I_\mu$, $c_\lambda(t) = c_\mu(t)$.

Proof. The fact that M is a submanifold of R^n and hence is Hausdorff is the key to the proof. See exercises for the details. ∎

10.2.3. Corollary. *The map* $c : I \to M$ *where* $I = \bigcup_{\lambda\in J} I_\lambda$ *defined by* $c(t) = c_\lambda(t)$ *for some* $\lambda \in J$ *is an integral curve for* Y *at* $a \in M$. ∎

10.2.4. Definition. Let Y, M, I and c be as in Lemma 10.2.2. and Corollary 10.2.3. The map $c : I \to M$ is called *the maximal integral curve for* Y *at* $a \in M$. ∎

EXERCISES 10.2.

1. *Let* M *be a smooth submanifold of* R^n, $Y : M \to TM$ *a smooth vectorfield and* $c : I \to M$ *a smooth parametrized curve. Show that* c *is an integral curve of* Y *if and only if, for each chart* (U,ϕ) *for* M, c_ϕ *is an integral curve of* Y_ϕ.

2. *Assume the hypotheses of Theorem 10.2.1. and suppose that for each interval* I *with* $0 \in I$ *there are distinct integral curves* $c : I \to M$ *and* $d : I \to M$ *for* Y *at* a. *Show that this contradicts Theorem 10.1.5.*

3. *Here we prove Lemma 10.2.2. Assume the hypotheses of the lemma. Let* I *be the subset of the interval* $I_\lambda \cap I_\mu$ *consisting of all* t *such that* $c_\lambda(t) = c_\mu(t)$. *The aim is to prove* $I = I_\lambda \cap I_\mu$. *This is done by showing:*

   (a) $I \neq \square$. *(Easy! Try* $t = 0$.*)*

   (b) I *is open in* $I_\lambda \cap I_\mu$. *First suppose* $s \in I$ *and set* $b = c_\lambda(s) = c_\mu(s)$. *Now consider an integral curve* $d : (-\varepsilon,\varepsilon) \to M$ *of* Y *at* b *where* $d(t) = c_\lambda(s+t)$ *for each* $t \in (-\varepsilon,\varepsilon)$. *Use Theorem 10.2.1 to show that for each* $t \in (-\varepsilon,\varepsilon)$ *we obtain* $c_\lambda(s+t) = c_\mu(s+t) = d(t)$, *which gives the required result.*

   (c) I *is closed in* $I_\lambda \cap I_\mu$. *(Use the fact that* M *is Hausdorff and the continuity of* c_λ *and* c_μ*).*

   (d) $I_\lambda \cap I_\mu = I$.

10.3. SECOND-ORDER VECTORFIELDS.

This section is about the special properties of the vectorfields which arise from the study of second-order differential equations, such as Lagrange's equations. In constructing a vectorfield from a second-order differential equation, we first apply the usual "reduction of order" procedure. The ideas of Section 10.1. can then be applied to the resulting pair of first-order equations.

10.3.1. Example. Given $f : R^k \times R^k \to R^k$, consider the second-order differential equation

$$\xi'' = f\circ(\xi,\xi') \tag{1}$$

for the unknown function $\xi : I \to R^k$. To reduce the order put $\eta = \xi'$ and thus get the pair of first-order equations

$$\begin{aligned} \xi' &= \eta \\ \eta' &= f\circ(\xi,\eta). \end{aligned} \tag{2}$$

These may be combined into the single first-order equation

$$(\xi,\eta)' = F\circ(\xi,\eta) \tag{3}$$

where $F : R^k \times R^k \to R^{2k}$ with $F(x,y) = (y, f(x,y))$.

To obtain the vectorfield Y corresponding to (3) we simply adjoin the "base point" (x,y) corresponding to (2) to the "right-hand side" $F(x,y)$, as in Section 10.1, to get

$$\begin{aligned} Y(x,y) &= ((x,y),\ F(x,y)) \\ &= ((x,y),\ (y,\ f(x,y)) \end{aligned} \tag{4}$$

Thus a vectorfield Y has been constructed in a natural way from the original second-order differential equation (2). ■

The following remarks concerning the above example are intended to motivate our definition below of a "second-order vectorfield".

10.3.2. Remarks. (a) The above vectorfield Y has the rather special property that, of the four components occurring on the right-hand side of (4), the second and third components are equal.

(b) If $(\xi,\mu): I \longrightarrow \mathrm{R}^k \times \mathrm{R}^k$ is any integral curve of the vectorfield Y then, in accordance with Section 10.1, it is a solution of the differential equation (3) and hence also of (2). This means in particular that

$$(\xi,\eta) = (\xi,\xi')$$

so that (ξ,η) is a self-consistent parametrized curve. ■

In generalizing these ideas to manifolds, we note that if a vectorfield $Y : M \longrightarrow TM$ is to have self-consistent integral curves then the manifold M should itself be the tangent bundle TQ of some submanifold Q of R^n. In the following definition, the elements of TQ and T^2Q are ordered pairs and ordered quadruples, respectively, of elements of R^n.

10.3.3. Definition. Let Q be a smooth submanifold of R^n. A vectorfield $Y : TQ \longrightarrow T^2Q$ is called a *second-order vectorfield* on TQ if for each $(x,y) \in TQ$ there is $z \in \mathrm{R}^n$ such that

$$Y(x,y) = ((x,y),\ (y,z)).\ ■$$

It is possible to express the idea that Y is a second-order vectorfield by using projection maps. This depends on the following lemma.

10.3.4. Lemma. *If* Q *is a submanifold of* R^n *and* $\tau_Q : TQ \longrightarrow Q$ *is the natural projection then the map*

$$T\tau_Q : T^2Q \longrightarrow TQ$$

has its value at $((a,b),(c,d)) \in T^2Q$ *given by*

$$T\tau_Q\,((a,b),(c,d)) = (a,c).$$

Proof. We use Definition 5.2.1. To this end, choose a differentiable parametrized curve γ in TQ at (a,b) with

$$\gamma'(0) = (c,d).$$

Now write γ componentwise as (γ_1,γ_2) with $\gamma_1 = \tau_Q \circ \gamma$ so that

$$(\tau_Q \circ \gamma)'(0) = \gamma_1'(0) = c.$$

Hence,

$$\begin{aligned} T\tau_Q((a,b),(c,d)) &= T\tau_Q((a,b),\ \gamma'(0)) \\ &= (\tau_Q(a,b),\ (\tau_Q \circ \gamma)'(0)) \\ &= (a,c). \quad \blacksquare \end{aligned}$$

The following corollary gives the desired characterization of a second-order vectorfield in terms of projections.

10.3.5. Corollary. *Let* Q *be a submanifold of* $\mathbb{R}^n$. *A vectorfield* $Y : TQ \to T^2Q$ *is second-order if and only if*

$$T\tau_Q \circ Y = \mathrm{id}_{TQ}\ .$$

Proof. Given a vectorfield $Y : TQ \to T^2Q$, let $(a,b) \in TQ$ and let

$$Y(a,b) = ((a,b),(c,d)) \in T^2Q\ .$$

By Lemma 10.3.4, the condition

$$(T\tau_Q \circ Y)(a,b) = (\mathrm{id}_{TQ})(a,b)$$

is equivalent to

$$(a,c) = (a,b),$$

and hence to

$$Y(a,b) = ((a,b),(b,d))\ .$$

This establishes the desired equivalence of the conditions in the corollary. ■

The next result shows that second-order vectorfields may be characterized in terms of their integral curves - a possibility which might be guessed from Example 10.3.1.

10.3.6. Theorem. *A smooth vectorfield* $Y : TQ \to T^2Q$ *is second-order if and only if each of its integral curves is self-consistent.*

Proof. Suppose first that $Y : TQ \to T^2Q$ is a smooth second-order vectorfield. Let $c: I \to TQ$ be an integral curve for Y so that

$$c^{;} = Y \circ c$$

and hence

$$\begin{aligned}(T\tau_Q) \circ c^{;} &= (T\tau_Q) \circ Y \circ c \\ &= id_{TQ} \circ c \qquad \text{by Corollary 10.3.5.} \\ &= c.\end{aligned}$$

It now follows from Exercise 10.3.3 that $(\tau_Q \circ c)^{;} = c$ and hence c is self-consistent.

Conversely, suppose that $Y : TQ \to T^2Q$ is a smooth vectorfield with the property that each of its integral curves is self-consistent. Let $(a,b) \in TQ$. By Theorem 10.2.1, there is an integral curve $c : I \to TQ$ of Y at (a,b). Hence

$$c^{;} = Y \circ c$$

and so

$$(T\tau_Q) \circ c^{;} = (T\tau_Q) \circ Y \circ c\ .$$

From Exercise 10.3.3 it now follows that

$$(\tau_Q \circ c)^{;} = (T\tau_Q) \circ Y \circ c$$

and since c is self-consistent this implies that

$$c = (T\tau_Q) \circ Y \circ c.$$

Evaluating both sides at 0 gives

$$(a,b) = (T\tau_Q) \circ Y(a,b).$$

Thus $T\tau_Q \circ Y = id_{TQ}$, showing that Y is second-order by Corollary 10.3.5. ∎

Thus the significant part of the integral curve c of a second-order vectorfield is the first component since the second is simply the derivative of the first.

10.3.7. Definition. If $c : I \to TQ$ is a (self-consistent) integral curve for a second-order vectorfield $Y : TQ \to T^2Q$ then $\tau_Q \circ c: I \to Q$ is called a *base integral curve of* Y. ∎

The following lemma will provide a useful tool for establishing a further criterion for second-order vectorfields, involving local representatives.

10.3.8. Lemma. *Let* Q *be a smooth submanifold of* R^n. *A parametrized curve* $c: I \to TQ$ *is self-consistent if and only if, for each chart* (U, ϕ) *for* Q, *the local representative* $c_{T\phi} = T\phi \circ c$ *is self-consistent.*

Proof. See the exercises. ■

10.3.9. Theorem. *Let* Q *be a smooth submanifold of* R^n. *A smooth vectorfield* $Y : TQ \to T^2Q$ *is second-order if and only if, for each chart* (U, ϕ) *for* Q, *the local representative* $Y_{T\phi}: TU \to T^2U$ *is second-order.*

Proof. Suppose that Y is second-order and let $\eta: I \to TQ$ be an integral curve of the vectorfield $Y_{T\phi}$. Put $c = T\phi^{-1} \circ \eta$ so that c is an integral curve of Y by Exercise 10.2.1. By Theorem 10.3.6, c is self-consistent. Hence, by Lemma 10.3.8, the local representative $c_{T\phi} = \eta$ is also self-consistent. It now follows from Theorem 10.3.6. that the vectorfield $Y_{T\phi}$ is second-order.

The proof in the converse direction is similar. ■

EXERCISES 10.3.

1. *Write down the vectorfield on* TR *corresponding to the following second-order differential equation for* $\xi : I \to R$:

$$\xi'' + 2\xi' + \xi = \underline{0}.$$

Find the maximal base integral curve of this vectorfield at $(a,b) \in TR$.

2. *Let* Q *be a submanifold of* R^n *and let* $Y : TQ \to T^2Q$ *be any map. Show that* Y *is a second-order vectorfield on* TQ *if and only if*

$$\tau_{TQ} \circ Y = id_{TQ} = T\tau_Q \circ Y$$

where $\tau_{TQ}: T^2Q \to TQ$ *and* $\tau_Q: TQ \to Q$ *are the natural projections.*

3. *Let* Q *be a submanifold of* R^n, *let* $c : I \to TQ$ *be a smooth parametrized curve and let* $\tau_Q: TQ \to Q$ *be the natural projection. Show from Definition 7.1.1 that*

$$(T\tau_Q) \circ c^{\cdot} = (\tau_Q \circ c)^{\cdot} .$$

4. *Let* Q *be a submanifold of* R^n *with* (U,ϕ) *as a chart and let* $\tau_Q : TQ \to Q$ *and* $\tau_{\phi(U)} : T\phi(U) \to \phi(U)$ *be the projections. Show from Definition 5.2.1 that, for each parametrized curve* $c : I \to TQ$,

$$\tau_{\phi(U)} \circ T\phi \circ c = \phi \circ \tau_Q \circ c .$$

5. *Prove Lemma 10.3.8. (Hint: the preceding exercise.)*

6. *Let* Q *be a submanifold of* R^n *and let* $Y : TQ \to T^2Q$ *be a smooth vectorfield. Show that* Y *is second-order if and only if each of its integral curves* $c : I \to TQ$ *can be written as* $c = d'$ *where* $d : I \to Q$ *satisfies* $d'' = Y \circ d'$.

11. LAGRANGIAN VECTORFIELDS

In this chapter the theory of vectorfields and their integral curves will be related back to the study of the motion of a particle on a submanifold Q *of* R^n. *Whereas in Section 8.3 we assumed the existence of a self-consistent differentiable curve* $c: I \to TQ$ *satisfying Newton's second law of motion, in this chapter one of the aims is to prove the existence of such curves - thereby verifying that the theory developed in Section 8.3 is not vacuous.*

To achieve this aim we shall use the idea that Lagrange's equations, with respect to each chart in an atlas for TQ, *define "local" vectorfields which can be patched together to form a "global" vectorfield on the whole of* TQ. *The existence-uniqueness theory of the preceding chapter can then be used to establish the existence of integral curves for this vectorfield. Finally, the integral curves so obtained can be shown to satisfy Newton's equations of motion.*

11.1. GLOBALIZING THEORY

This section sets up some language which is useful for studying how local vectorfields may be patched together to form a global vectorfield on a manifold. Throughout this section M will denote a k-dimensional submanifold of R^n.

11.1.1. Definition. Let (U,ϕ) and (V,ψ) be two compatible charts for M. A pair of local vectorfields

$$Y : \phi(U) \to T\phi(U) \quad \text{and} \quad Z : \psi(V) \to T\psi(V)$$

is said to be *patchable* with respect to these charts if, on the domain $U \cap V$,

$$(T\phi)^{-1} \circ Y \circ \phi = (T\psi)^{-1} \circ Z \circ \psi. \quad \blacksquare$$

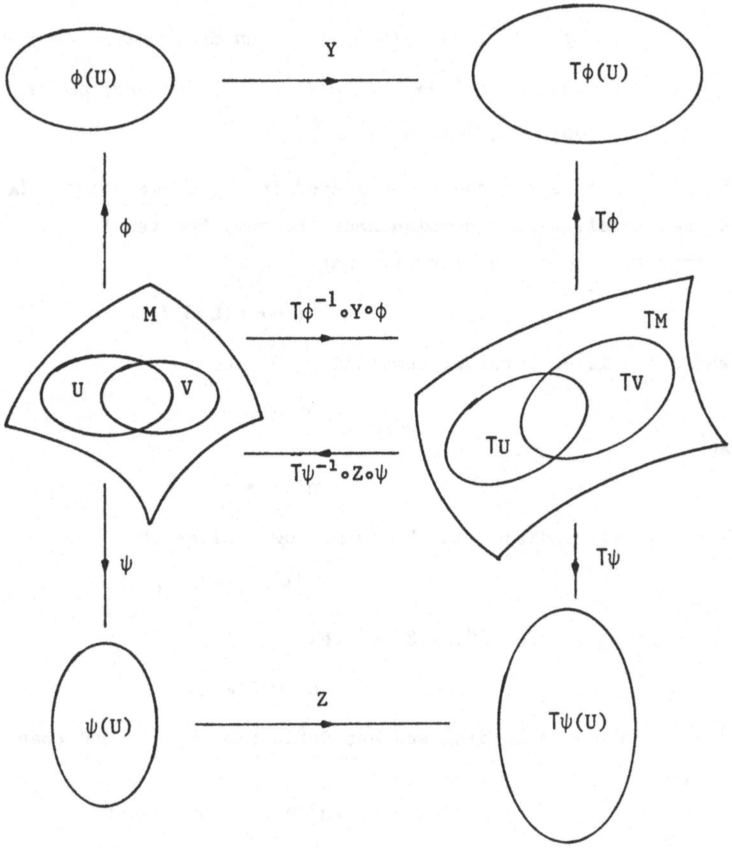

Figure 11.1.1. Patchable local vectorfields.

The motivation for this definition is that Y and Z will satisfy it whenever there is a global vectorfield on the manifold M of which Y and Z are the local representatives with respect to the charts (U,ϕ) and (V,ψ) as is illustrated in Figure 11.1.1.

In applications to specific examples it will sometimes be possible to avoid awkward computations by using the following theorem - which shifts attention away from the vectorfields to their integral curves.

11.1.2. Lemma. *Let* (U,ϕ) *and* (V,ψ) *be two compatible charts for* M. *A pair of local vectorfields*

$$Y : \phi(U) \to T\phi(U) \quad \textit{and} \quad Z : \psi(V) \to T\psi(V)$$

is patchable with respect to these charts if the following condition holds:

if $\eta : I \to \phi(U \cap V)$ *is an integral curve for* Y

then $\zeta : I \to \psi(U \cap V)$ *is an integral curve for* Z

where $\zeta = (\psi \circ \phi^{-1}) \circ \eta$.

Proof. Suppose the condition stated in the lemma holds. Let $a \in \phi(U \cap V)$. By the local existence-uniqueness theorem, Theorem 10.1.5, there is an integral curve at a for Y, say

$$\eta : I \to \phi(U \cap V)$$

where I is an interval containing 0. Hence

$$\eta^{;} = Y \circ \eta \tag{1}$$

and

$$\eta(0) = a. \tag{2}$$

The assumed condition in the lemma now implies that

$$\zeta = (\psi \circ \phi^{-1}) \circ \eta \tag{3}$$

is an integral curve for Z so that

$$\zeta^{;} = Z \circ \zeta. \tag{4}$$

Now substitute (3) into (4) and use Definition 7.1.1 and then the chain rule to get

$$T(\psi \circ \phi^{-1}) \circ \eta^{;} = Z \circ (\psi \circ \phi^{-1}) \circ \eta.$$

Hence

$$T(\psi \circ \phi^{-1}) \circ Y \circ \eta = Z \circ (\psi \circ \phi^{-1}) \circ \eta \qquad \text{by (1)}$$

$$T(\psi \circ \phi^{-1}) \circ Y(a) = Z \circ (\psi \circ \phi^{-1})(a) \qquad \text{by (2)}$$

and so

$$T\phi^{-1} \circ Y(a) = T\psi^{-1} \circ Z \circ \psi \circ \phi^{-1}(a).$$

But as a is an arbitrary element of $\phi(U \cap V)$ this implies that, on the domain $\phi(U \cap V)$,

$$T\phi^{-1} \circ Y = T\psi^{-1} \circ Z \circ \psi \circ \phi^{-1}$$

and hence, on the domain $U \cap V$,

$$T\phi^{-1} \circ Y \circ \phi = T\psi^{-1} \circ Z \circ \psi. \quad \blacksquare$$

The next theorem is a trivial consequence of our definition of patchability.

11.1.3. Theorem. *Let* $\{(U^\lambda,\phi^\lambda): \lambda \in J\}$ *be an atlas for* M *and, for each* $\lambda \in J$, *let* Y^λ *be a smooth local vectorfield on* $\phi(U^\lambda)$. *Suppose that, for each* $\lambda, \mu \in J$, *the vectorfields* Y^λ *and* Y^μ *are patchable with respect to the charts* $(U^\lambda, \phi^\lambda)$ *and* (U^μ, ϕ^μ). *There is then a unique smooth vectorfield* Y *on* M *such that each* Y^λ *is the local representative of* Y *in the chart* $(U^\lambda, \phi^\lambda)$.

Proof. We may define Y on each U^λ by putting

$$Y|U^\lambda = (T\phi^\lambda)^{-1} \circ Y^\lambda \circ \phi^\lambda .$$

This definition is unambiguous by the definition of patchability. This, moreover, is the only possible choice of Y. ■

11.1.4. Corollary. *If in Theorem 11.1.3 the manifold* M *is the tangent bundle of some other manifold and each local vectorfield* Y^λ *is second-order then the vectorfield* Y *on* M *is also second-order.*

Proof. This is an immediate consequence of Theorem 10.3.9. ■

11.2. APPLICATION TO LAGRANGE'S EQUATIONS

The study of the motion of a particle on a manifold leads, via Lagrange's equations with respect to the charts on the manifold, to local second-order vectorfields. In this section it will be shown that these vectorfields can be patched together to give a vectorfield defined on the whole of the tangent bundle of the manifold.

In order to be able to apply the globalizing theory of Section 11.1, we must first check that the local vectorfields obtained from Lagrange's equations are patchable. For this task, the next lemma is crucial. Note that the function L which occurs in the lemma is not restricted to being the Lagrangian for any particle motion. Birkhoff(1927), page 21, has an alternative albeit more coordinatewise proof.

Throughout this section, Q will denote a k-dimensional submanifold of R^n.

11.2.1. Lemma. *Consider any smooth function* $L : TQ \to R$. *If* (U, q) *and* (V, r) *are charts for* Q *and if* $c : I \to U \cap V$ *is any self-consistent smooth curve then*

$$\left(\frac{\partial L}{\partial \dot{q}} \circ c\right)' - \frac{\partial L}{\partial q} \circ c = \underline{0} \Rightarrow \left(\frac{\partial L}{\partial \dot{r}} \circ c\right)' - \frac{\partial L}{\partial r} \circ c = \underline{0} .$$

Proof. A mild extension of Corollary 7.2.8, shows that

$$\frac{\partial L}{\partial q} = \frac{\partial L}{\partial \dot{r}}\frac{\partial \dot{r}}{\partial q} + \frac{\partial L}{\partial r}\frac{\partial r}{\partial q} \circ \tau_Q \tag{1}$$

and also

$$\frac{\partial L}{\partial \dot{q}} = \frac{\partial L}{\partial \dot{r}}\frac{\partial \dot{r}}{\partial \dot{q}} \tag{2}$$

since r depends only on q so that $\frac{\partial r}{\partial \dot{q}} = \underline{0}$. We also have by Corollary 7.3.4,

$$\frac{\partial \dot{r}}{\partial \dot{q}} \circ c = \frac{\partial r}{\partial q} \circ \tau_Q \circ c \tag{3}$$

and hence by Corollary 7.3.8.

$$\left(\frac{\partial \dot{r}}{\partial \dot{q}} \circ c\right)' = \frac{\partial \dot{r}}{\partial q} \circ c\ . \tag{4}$$

Now suppose, in accordance with the hypotheses of the theorem, that

$$\left(\frac{\partial L}{\partial \dot{q}} \circ c\right)' - \frac{\partial L}{\partial q} \circ c = \underline{0} \tag{5}$$

where by (2)

$$\left(\frac{\partial L}{\partial \dot{q}} \circ c\right)' = \left(\frac{\partial L}{\partial \dot{r}} \circ c\right)' \frac{\partial \dot{r}}{\partial \dot{q}} \circ c + \frac{\partial L}{\partial \dot{r}} \circ c \left(\frac{\partial \dot{r}}{\partial \dot{q}} \circ c\right)'$$

$$= \left(\frac{\partial L}{\partial \dot{r}} \circ c\right)' \frac{\partial r}{\partial q} \circ \tau_Q \circ c + \frac{\partial L}{\partial \dot{r}} \circ c\ \frac{\partial \dot{r}}{\partial q} \circ c$$

by (3) and (4)

while by (1)

$$\frac{\partial L}{\partial q} \circ c = \frac{\partial L}{\partial \dot{r}} \circ c\ \frac{\partial \dot{r}}{\partial q} \circ c + \frac{\partial L}{\partial r} \circ c\ \frac{\partial r}{\partial q} \circ \tau_Q \circ c\ .$$

Hence (5) becomes

$$\left(\left(\frac{\partial L}{\partial \dot{r}} \circ c\right)' - \frac{\partial L}{\partial r} \circ c\right) \frac{\partial r}{\partial q} \circ \tau_Q \circ c = 0.$$

But as (U,q) and (V,r) are both charts for Q, it follows that the matrix on the right is non-singular and hence

$$\left(\frac{\partial L}{\partial \dot{r}} \circ c\right)' - \frac{\partial L}{\partial r} \circ c = \underline{0}. \quad \blacksquare$$

Our next result recasts Lemma 11.2.1 into a form which refers to local representatives of curves rather than to the curves themselves.

Motivation for this comes from Chapter 9 where we studied the explicit form of Lagrange's equations for the local representatives of the curves representing the history of the particle. Recall that if $c: I \to TQ$ is a curve and (TU, Tq) is a natural chart for TQ then the local representative c_{Tq} of c is defined by

$$c_{Tq} = Tq \circ c \quad \text{or} \quad c = (Tq)^{-1} \circ c_{Tq} .$$

In addition to the notation $(q \circ \tau_Q \circ c,\ \dot{q} \circ c)$ for the local representative of c, we also used the notation $(\bar{q}, \bar{q}')$ to give our answers in Chapter 9 a more customary appearance.

Since, however, we now wish to change tack and take the local representatives themselves as our starting point, we shall use letters such as "ξ" and "η" to denote them. In this way we avoid the possibility of begging the question as to whether we can construct a suitable global curve $c : I \to TQ$ which is independent of charts.

11.2.2. **Lemma.** *Consider any smooth function* $L : TQ \to R$. *Let* (U,q) *and* (V,r) *be charts for* Q *and let*

$$\eta: I \to (Tq)(U \cap V) \quad \textit{and} \quad \zeta: I \to (Tr)(U \cap V)$$

be any self-consistent smooth curves.

$$\textit{If} \qquad \left(\frac{\partial L}{\partial \dot{q}} \circ (Tq)^{-1} \circ \eta\right)' - \frac{\partial L}{\partial q} \circ (Tq)^{-1} \circ \eta = \underline{0} \tag{1}$$

$$\textit{then} \qquad \left(\frac{\partial L}{\partial \dot{r}} \circ (Tr)^{-1} \circ \zeta\right)' - \frac{\partial L}{\partial r} \circ (Tr)^{-1} \circ \zeta = \underline{0} \tag{2}$$

$$\textit{where} \qquad \zeta = Tr \circ (Tq)^{-1} \circ \eta . \tag{3}$$

Proof. Define a parameterized curve $c: I \to U \cap V$ by putting

$$c = (Tq)^{-1} \circ \eta . \tag{4}$$

From the hypothesis (1) of the lemma

$$\left(\frac{\partial L}{\partial \dot{q}} \circ c\right)' - \frac{\partial L}{\partial q} \circ c = \underline{0} .$$

Since η is self-consistent, however, the same is true of c, by Lemma 10.3.8. Hence, by Lemma 11.2.1,

$$\left(\frac{\partial L}{\partial \dot{r}} \circ c\right)' - \frac{\partial L}{\partial r} \circ c = \underline{0}. \tag{5}$$

But from (4) and the hypothesis (3) in the lemma,

$$c = (Tr)^{-1} \circ \zeta.$$

Substituting this in (5) gives the desired conclusion (2) of the lemma. ∎

The stage has now been set for an application of the globalizing theory of Section 11.1 to the local vectorfields defined by Lagrange's equations. The assumption that the function $L : TQ \to R$ is the Lagrangian of a mechanical system will now be used for the first time in this section.

11.2.3. Theorem. *Let* $L : TQ \to R$ *be the Lagrangian of the particle moving on the submanifold* Q *as described in Section 8.3. If* $\{(U^\lambda, q^\lambda) : \lambda \in J\}$ *is an atlas for* Q *then*

(a) for each $\lambda \in J$ *there is a smooth local second-order vectorfield*

$$Y^\lambda : Tq^\lambda(U^\lambda) \to T^2q^\lambda(U^\lambda)$$

whose integral curves are the solutions η^λ *of Lagrange's equations*

$$\left(\frac{\partial L}{\partial \dot{q}^\lambda} \circ (Tq^\lambda)^{-1} \circ \eta^\lambda\right)' - \frac{\partial L}{\partial q^\lambda} \circ (Tq^\lambda)^{-1} \circ \eta^\lambda = \underline{0} ,$$

(b) there is a unique smooth second-order vectorfield

$$Y : TQ \to T^2Q$$

for which the local representative in the chart (TU^λ, Tq^λ) *is* Y^λ .

Proof. By Theorem 9.3.3, for each $\lambda \in J$, Lagrange's equations in the form written above are second-order differential equations for the function $\bar{q}^\lambda$ where $\eta^\lambda = (\bar{q}^\lambda, \bar{q}^{\lambda\prime})$. Thus, as in Section 10.3, they define a smooth local second-order vectorfield Y^λ with the stated integral curves.

By Lemmas 11.2.2 and 11.1.2, each pair of vectorfields Y^λ and Y^μ is locally patchable with respect to the charts (TU^λ, Tq^λ) and (TU^μ, Tq^μ), so Theorem 11.1.3 gives the existence of the desired vectorfield Y. ∎

The vectorfield Y occurring in Theorem 11.2.3 will be called the Lagrangian vectorfield of the mechanical system described in Section 8.3.

11.2.4. Corollary. *If* Y *is the vectorfield given by Theorem 11.2.3 then for each* $\dot{a} \in TQ$ *there is a unique integral curve* $c : I \to TQ$ *for* Y *at* $\dot{a}$ *having maximal domain. This curve is self-consistent and for any* *chart* (U, q) *for* Q *it satisfies Lagrange's equations*

$$\left(\frac{\partial L}{\partial \dot{q}} \circ c\right)' - \frac{\partial L}{\partial q} \circ c = \underline{0} \ .$$

Proof. Left for the exercises. ∎

EXERCISE 11.2.

Prove Corollary 11.2.4.

11.3. BACK TO NEWTON.

It will now be shown that the arguments leading from Newton's second law of motion to Lagrange's equations can be reversed. A result in this reverse direction is given in Landau & Liftshitz(1960), page 9, although their result applies only in the more rudimentary context in which there are no geometric constraints on the motion. The following lemma will be used in the proof of our result.

11.3.1. Lemma. *Let* X *be the identity map on* R^n, *let* Q *be a submanifold of* R^n *of dimension* k *and let* $a \in Q$. *If* (U,q) *is a chart for* Q *at* a *then the set of vectors*

$$\left(\frac{\partial X}{\partial q_1}(a), \frac{\partial X}{\partial q_2}(a), \ldots, \frac{\partial X}{\partial q_k}(a)\right)$$

is a basis for T_aQ.

Proof. For $1 \leqslant i \leqslant k$ define $\gamma_i : I \longrightarrow U$ by putting

$$\gamma_i(t) = q^{-1}(q(a) + t\, e_i)$$

so that, as in the proof of Theorem 7.3.10,

$$\gamma_i{}'(0) = \frac{\partial X}{\partial q_i}(a) \ .$$

But by the construction of γ_i

$$q \circ \gamma_i = \underline{q(a)} + id_I\, e_i$$

and hence by the chain rule

$$Tq \circ T\gamma_i = T(\underline{q(a)} + id_I\, e_i)$$

$$Tq(a, \gamma_i{}'(0)) = (q(a), e_i).$$

But since $Tq|T_aQ$ is linear and the set of vectors $(e_1, e_2, \ldots, e_k)$ is linearly independent the same is true of the set of vectors $(\gamma_1'(0), \gamma_2'(0),\ldots,\gamma_k'(0))$, which therefore is a basis for the k-dimensional vector space T_aQ. ■

Although our theorem is stated and proved only in the context of Section 8.2, where the submanifolds lie in R^3, it can easily be extended to the more general situation discussed in Section 8.3.

11.3.2. Theorem. *Consider a particle of mass* m *moving on a k-dimensional submanifold* Q *of* R^3 *subject to a conservative field of force. Let* $T : TQ \to R$ *and* $V : Q \to R$ *be the kinetic and potential energies of the particle and let* $L = T - V\circ\tau_Q$ *be the Lagrangian.*

The reaction force R*, constraining the particle to stay on the manifold, can then be chosen in just one way at each point of* TQ *in order that the following condition holds:*

if $c: I \to TQ$ *is an integral curve for the Lagrangian vectorfield on* TQ *determined by* L *then Newton's second law holds for each* $t \in I$:

$$m(X\circ\tau_Q\circ c)''(t) = -\left(\frac{\partial V}{\partial X}\circ\tau_Q + \pi_2\circ R\right)\circ c(t)\ .$$

At each point the reaction force R *is orthogonal to* Q.

Proof. Let $c : I \to TQ$ be an integral curve for the Lagrangian vectorfield so that, as in Corollary 11.2.4, c is self-consistent. Let $t \in I$ and let (U,q) be a chart for Q such that $c(t) \in U$. The argument contained in the proof of Theorem 8.2.2 which shows that for $1 \leq i \leq k$

$$\left(\frac{\partial T}{\partial \dot{q}_i}\circ c\right)' - \frac{\partial T}{\partial q_i}\circ c = m(\dot{X}\circ c)'\cdot\left(\frac{\partial X}{\partial q_i}\circ\tau_Q\circ c\right) \tag{1}$$

remains valid. By the hypotheses of the theorem, moreover,

$$\left(\frac{\partial L}{\partial \dot{q}_i}\circ c\right)' - \frac{\partial L}{\partial q_i}\circ c = \underline{0}$$

and since $L = T - V\circ\tau_Q$ this implies that

$$\left(\frac{\partial T}{\partial \dot{q}_i}\circ c\right)' - \frac{\partial T}{\partial q_i}\circ c + \frac{\partial V}{\partial q_i}\circ\tau_Q\circ c = \underline{0}\ . \tag{2}$$

Subtracting (2) from (1) now gives

$$m(\dot{X}\circ c)'\cdot\left(\frac{\partial X}{\partial q_i}\circ\tau_Q\circ c\right) = -\frac{\partial V}{\partial q_i}\circ\tau_Q\circ c$$

$$= -\left(\frac{\partial V}{\partial X}\frac{\partial X}{\partial q_i}\right)\circ \tau_Q \circ c$$

so that

$$\left(m(\dot{X}\circ c)' + \frac{\partial V}{\partial X}\circ \tau_Q \circ c\right)\cdot\frac{\partial X}{\partial q_i}\circ \tau_Q \circ c = \underline{0}\ .$$

But since this holds for $1 \leqslant i \leqslant k$ we may apply Lemma 11.3.1 after evaluating both sides at t, to deduce that the vector

$$m(\dot{X}\circ c)'(t) + \frac{\partial V}{\partial X}\circ \tau_Q \circ c(t)$$

is orthogonal to the tangent space $T_{c(t)}Q$. Hence there is a function $R : TQ \to TQ$ such that

$$m(\dot{X}\circ c)'(t) + \frac{\partial V}{\partial X}\circ \tau_Q \circ c(t) = \pi_2 \circ R \circ c(t)$$

and $R(c(t))$ is orthogonal to $T_{c(t)}Q$. ■

EXERCISE 11.3.

Extend Theorem 11.3.2 to the context of Section 8.3 and then prove it.

12. FLOWS

At any particular time the state of a system is given by some collection of parameters. A deterministic mathematical model would hope to predict future states of that system. So if say a *is the state at time* $t = 0$ *we may have a state* $\Lambda^t(a)$ *at time* t. *After another time lapse of say* s *we would have a state* $\Lambda^s(\Lambda^t(a))$. *We would therefore want our model to satisfy the relation* $\Lambda^s(\Lambda^t(a)) = \Lambda^{t+s}(a)$, *that is,* $\Lambda^{t+s} = \Lambda^s \circ \Lambda^t$. *In other words the "time evolution" of our system would have to satisfy certain group properties. We could thus think of the "time evolution" operator* Λ^t *as acting on all points of a manifold of states at once and think of this as giving a "flow" on the manifold with increasing time.*

In the case of a mechanical system our manifold is TQ *and the flow is a collection of integral curves in* TQ *of a Lagrangian vectorfield, each such curve corresponding to a set of initial conditions.*

12.1. FLOWS GENERATED BY VECTORFIELDS

Given a smooth vectorfield on a manifold, it was shown in Chapter 10 that there is a unique integral curve at each point of the manifold. Here our concern is with the overall structure of the set of all such integral curves obtained as the point ranges over the manifold.

It is helpful to regard the integral curves as the paths traced out by particles of a fluid moving with velocities which are prescribed at each point by the vectorfield. In this sense the vectorfield is said to determine a "flow" on the manifold. Intuitively, we may think of the flow as the set of all the integral curves of the vectorfield. The formal definition given later, however, expresses the idea of a flow in terms of certain mappings associated with the integral curves. An example will be given below to show how these mappings arise and to illustrate their special properties.

In the study of flows, attention is usually restricted to vectorfields which satisfy the following definition.

12.1.1. Definition. A smooth vectorfield Y on a manifold M is said to be *complete* if every integral curve for Y has the whole of R as its domain. ■

Not all vectorfields have this property (as Exercise 10.1.1. shows) but it can often be shown to hold for the vectorfields arising in Lagrangian mechanics.

12.1.2. Example. The vectorfield on R^2 of Example 10.1.2. given by

$$Y(x,y) = ((x,y),(-y,x))$$

has at each point $(a,b) \in R^2$ the integral curve $c: R \to R^2$ with

$$c(t) = (a \cos(t) - b \sin(t),\ a \sin(t) + b \cos(t)) \qquad (1)$$

Thus each integral curve is defined on the whole of R and so the vectorfield is complete.

With a change of viewpoint we may regard $c(t)$, as given by (1), as function of the point $(a,b) \in R^2$. This determines a map $\Lambda^t : R^2 \to R^2$ for each $t \in R$ with

$$\Lambda^t(a,b) = (a \cos(t) - b \sin(t),\ a \sin(t) + b \cos(t))$$

This map Λ^t can be regarded as a "time evolution" operator which assigns to each initial state of the system its final state after the lapse of a time t. Thus Λ^t is a map which acts on the whole of R^2, rotating each point around the origin anticlockwise through an angle t. Hence it is clear that

$$\Lambda^{t+s}(a,b) = \Lambda^t(\Lambda^s(a,b))$$
$$= \Lambda^t \circ \Lambda^s(a,b)$$

or, more simply,

$$\Lambda^{t+s} = \Lambda^t \circ \Lambda^s \qquad (2)$$

Intuitively, this means that the time evolution operator can be applied to initial conditions (a,b) given at an arbitrarily assigned initial instant. The property (2) is illustrated in Figure 12.1.1.

Other properties of the maps $\{\Lambda^t : t \in R\}$ are

$$\Lambda^0 = id_{R^2}$$
$$\Lambda^t \circ \Lambda^{-t} = id_{R^2}$$

so Λ^{-t} can be regarded as the operator which moves back along the integral curves for a time t. Most importantly, for each $t \in \mathbb{R}$,

$$\Lambda^t : \mathbb{R}^2 \to \mathbb{R}^2$$

is a diffeomorphism. ■

In order to set the above properties of the time-evolution operators in a natural algebraic setting, it is appropriate at this stage to introduce some notation for the set of diffeomorphisms from a manifold into itself.

12.1.3. Definition. Let M be a manifold. We denote by $\mathrm{Diff}^\infty(\mathrm{M})$ *the set of all* $\mathfrak{C}^\infty$ *diffeomorphisms from* M *to* M. ■

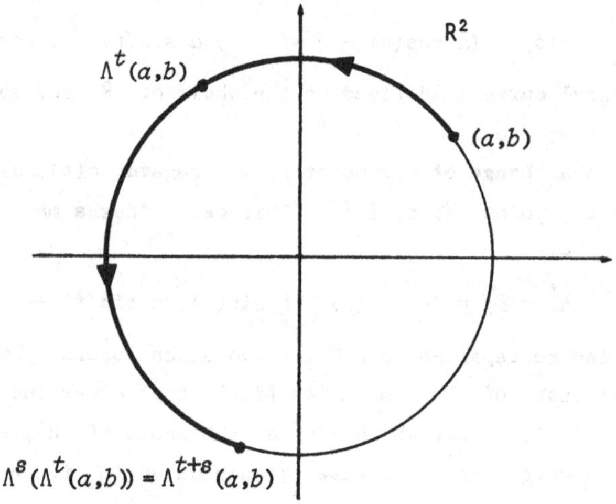

Figure 12.1.1.

12.1.4. Example. In each case a subset of the given set of diffeomorphisms is shown:

(a) $\{\lambda\, \mathrm{id}_{\mathbb{R}^k} : \lambda \in \mathbb{R} \text{ and } \lambda \neq 0\} \subseteq \mathrm{Diff}^\infty(\mathbb{R}^k)$

(b) $\{\mathrm{id}_{\mathbb{R}^k} + \underline{c} : c \in \mathbb{R}^k\} \subseteq \mathrm{Diff}^\infty(\mathbb{R}^k)$

(c) $\{\mathrm{id}_{\mathbb{R}^2} + t\,\underline{c} : t \in \mathbb{R},\ c \in \mathbb{R}^2\} \subseteq \mathrm{Diff}^\infty(\mathbb{R}^2)$

(d) $\{e^t\, \mathrm{id} : t \in \mathbb{R}\} \subseteq \mathrm{Diff}^\infty(\mathbb{R})$. ■

12.1.5. Lemma. *The set* $\mathrm{Diff}^\infty(\mathrm{M})$ *is a group under the operation of composition.* ■

The property (2) of the time evolution operators in Example 12.1.2. can now be expressed in terms of group theory. Since the map

$$t \to \Lambda^t$$

has the group $(R,+)$ as its domain and the group $(\mathrm{Diff}(R^2), \circ)$ as its codomain, property (2) simply says that this map preserves the group structure and hence is a *group homomorphism*.

The formal definition of a flow can now be given. It makes no direct reference to vectorfields, although they provide the motivation for introducing the concept.

12.1.6. Definition. A *flow* on a manifold M is a map

$$\Lambda : R \longrightarrow \mathrm{Diff}^{\infty}(M)$$

such that, for each s and $t \in R$, $\Lambda(t+s) = \Lambda(t)\circ\Lambda(s)$. We also require that the map $(a,t) \mapsto \Lambda(t)(a)$ be smooth. ■

Thus the image $\Lambda(R)$ of R under a flow Λ is a copy of the real line in $\mathrm{Diff}^{\infty}(M)$ with the same group properties as R and is sometimes called a *one-parameter group* of diffeomorphisms.

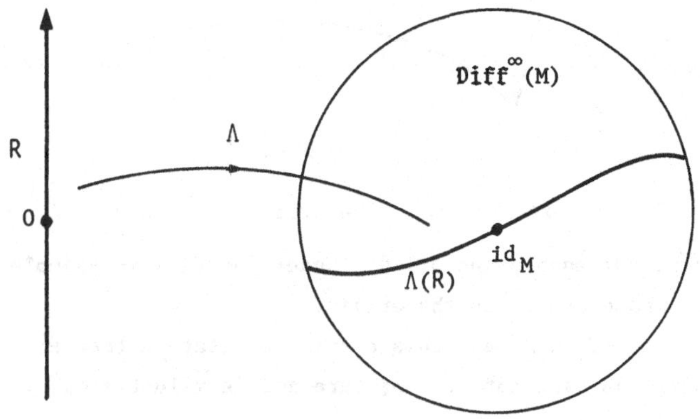

Figure 12.1.2.

12.1.7. Examples. The following maps are flows on the manifold R:

(a) $\Lambda: R \to \mathrm{Diff}^{\infty}(R) : t \mapsto e^t\,\mathrm{id}$

(b) $\Lambda: \mathrm{R} \longrightarrow \mathrm{Diff}^{\infty}(\mathrm{R}^2)$

$$: t \longmapsto (\cos(t)\mathrm{id}_1 - \sin(t)\mathrm{id}_2, \quad \sin(t)\mathrm{id}_1 + \cos(t)\mathrm{id}_2). \quad \blacksquare$$

The following definition recovers the idea of an integral curve, or more precisely its image, from the flow.

12.1.8. Definition. Let Λ be a flow on a manifold M. The *orbit* of a point $a \in \mathrm{M}$ under the flow Λ is the set of points from M

$$\{\Lambda(t)(a) : t \in \mathrm{R}\}. \quad \blacksquare$$

This idea is illustrated in Figure 12.1.3.

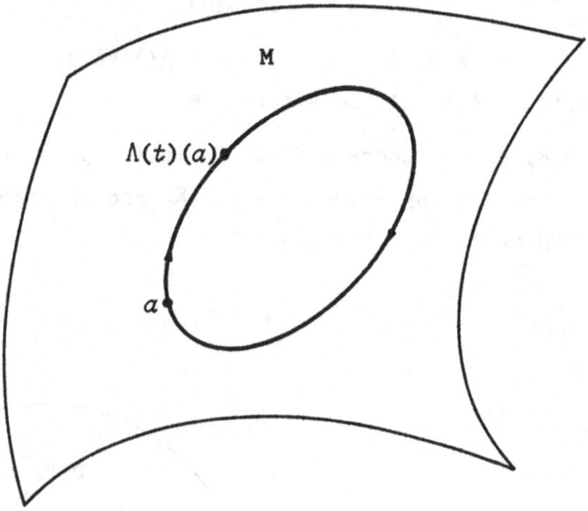

Figure 12.1.3. The orbit of a under Λ

The orbit of each point of R^2 under the flow in Example 12.1.7(b) is a circle whose centre is the origin.

To complete this section we state a theorem which brings vectorfields back into the picture and is illustrated by Example 12.1.2.

12.1.9. Theorem. *Let* Y *be a smooth vectorfield on a manifold* M. *If* Y *is complete then putting*

$$\Lambda(t)(a) = \mathrm{c}(t),$$

where c *is the maximal integral curve of* Y *at* a, *defines a flow on* M.

Proof. This involves ideas and results from Abraham & Marsden(1978), Section 2.1. $\blacksquare$

EXERCISES 12.1.

1. *Check that each of the subsets in Example 12.1.4. is a subgroup of the indicated group of diffeomorphisms, under composition.*

2. *Let* $\Lambda : \mathrm{R} \longrightarrow \mathrm{Diff}^{\infty}(\mathrm{M})$ *be a flow on a manifold* M. *Show that* $\Lambda(0) = \mathrm{id}_{\mathrm{M}}$ *and, for each* $t \in \mathrm{R}$, $(\Lambda(t))^{-1} = \Lambda(-t)$.

3. *Verify that the map* Λ *of Example 12.1.7(a) is a flow on* R. *Write down its one-parameter group of diffeomorphisms. Give the orbit of the point* $a \in \mathrm{R}$ *in each of the cases* $a < 0$, $a = 0$ *and* $a > 0$.

4. *Verify that the map* $\Lambda : \mathrm{R} \longrightarrow \mathrm{Diff}^{\infty}(\mathrm{TR})$ with

$$\Lambda(t)(a,v) = \left(\frac{v}{k}\sin(kt) + a\cos(kt),\ v\cos(kt) - ak\sin(kt)\right)$$

where $k > 0$, *is a flow on* TR. *Sketch some typical orbits.*

12.2. FLOWS FROM MECHANICS

Some familiar problems from mechanics will be used here to illustrate the idea of the flow of a vectorfield. The theoretical background for these examples is very straightforward. The vectorfields which arise in these problems can be proved complete by an application of the results in the next section; the integral curves of these vectorfields then generate flows in accordance with Theorem 12.1.9. Hence our discussion will concentrate on the geometrical ideas associated with the flows.

Recall that a flow $\Lambda : \mathrm{R} \longrightarrow \mathrm{Diff}^{\infty}(\mathrm{M})$ generated by a complete vectorfield Y on a manifold M determines two families of mappings

(a) for each $t \in \mathrm{R}$, the *time evolution operator* for lapse of time t

$$\Lambda^{t} : \mathrm{M} \longrightarrow \mathrm{M} : a \longrightarrow \Lambda(t)(a)$$

(b) for each $a \in \mathrm{M}$, the *integral curve* at a

$$\Lambda_{a} : \mathrm{R} \longrightarrow \mathrm{M} : t \longrightarrow \Lambda(t)(a).$$

In addition, there is for each $a \in \mathrm{M}$ the subset of M consisting of the image of the integral curve through a and called the *orbit* of the flow through a.

In the mechanical problems which follow, the configuration manifolds are 1-dimensional and hence their tangent bundles, which contain the orbits, are 2-dimensional. This permits direct diagrammatic representation of the orbits and also, but less directly, of the integral curves and the time evolution operators. Thus in Figure 12.2.1, the curves with arrows are the orbits of a flow Λ. From a knowledge of the times at which the particle assumes the various states, it is possible to infer what Λ^t is doing.

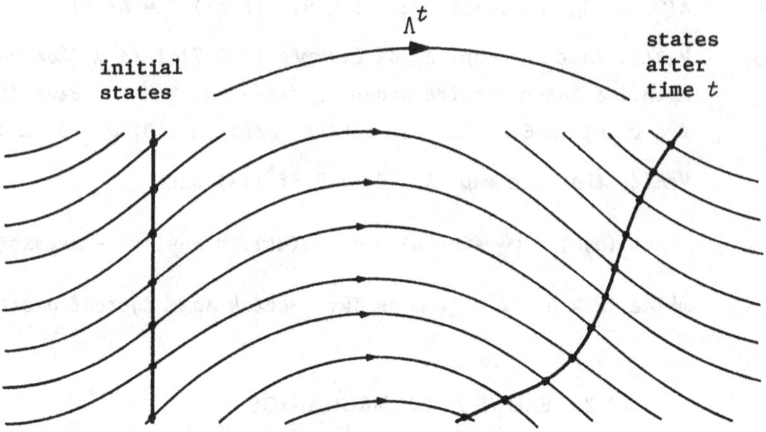

Figure 12.2.1. Time evolution operator.

12.2.1. Example. (Motion under gravity in one dimension.) For a particle of mass m moving under gravity and with the x-coordinate measuring the height, the Lagrangian L is given by

$$L = \tfrac{1}{2} m\dot{x}^2 - mg$$

and hence Lagrange's equation becomes

$$(x \circ c)'' = -g .$$

When expressed as a vectorfield, as in Example 10.3.1, this gives $Y : \mathsf{R} \to \mathsf{TR}$ with

$$Y(x,w) = ((x,w),(w,-g))$$

for each $(x,w) \in \mathsf{TR}$. The vectorfield Y is shown in Figure 12.2.2.

Elementary differential equations techniques show that the integral curve for Y at the point $(a,v) \in \mathsf{TR}$ is given by

$$c(t) = (a + vt - \tfrac{1}{2}gt^2,\ v-gt)$$

and hence the flow $\Lambda : \mathsf{R} \to \mathrm{Diff}^\infty(\mathsf{TR})$ is given by

$$\Lambda(t)(a,v) = (a+vt - \tfrac{1}{2}gt^2,\ v-gt)$$

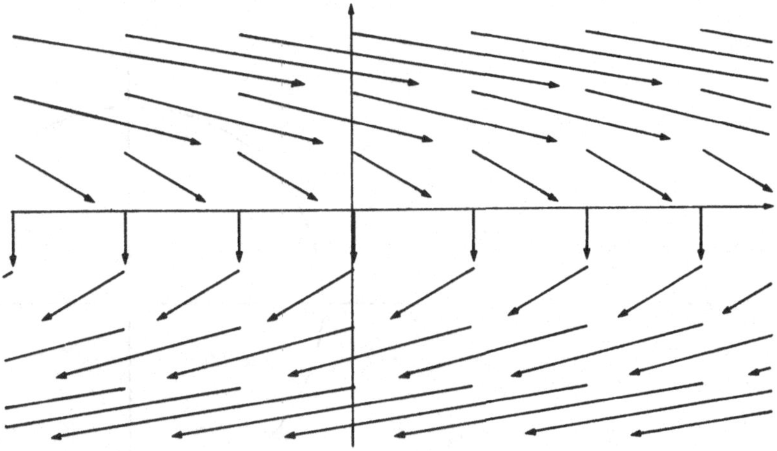

Figure 12.2.2. Gravitational vectorfield on phase space.

Note that each integral curve is self-consistent and that the orbits consist of parabolas which are tangent to the vectorfield. See Figure 12.2.3.

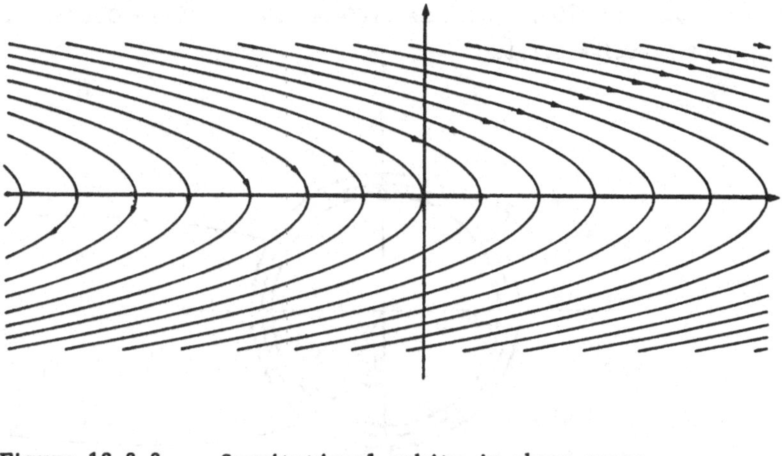

Figure 12.2.3. Gravitational orbits in phase space.

12.2.2. **Example.** (The simple pendulum.) This consists of a particle of mass m constrained to move under gravity on a circle lying in a vertical plane, in the absence of friction.

Distances in the horizontal and vertical directions will be measured by x and y-coordinates, respectively, and the circle on which the particle moves with be taken as S^1. See Figure 12.2.4.

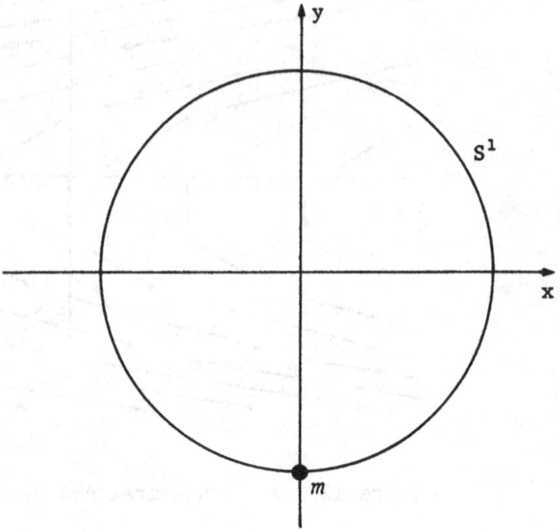

Figure 12.2.4. A simple pendulum

Thus the configuration manifold for the particle is S^1 and the velocity phase space is TS^1, which is diffeomorphic to the cylinder $S^1 \times R$ in view of Exercise 6.1.1.

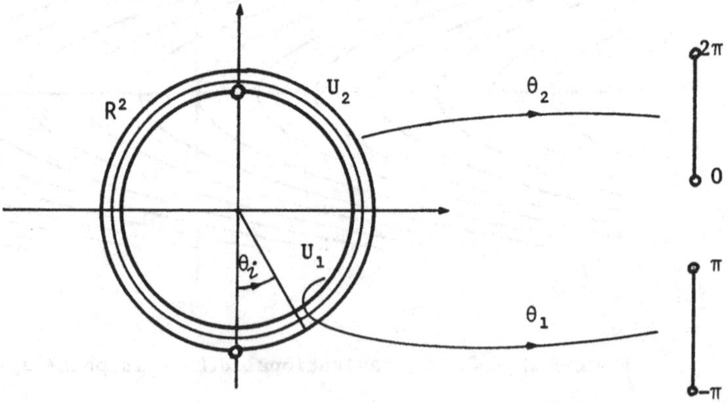

Figure 12.2.5. Charts for S^1 .

We shall use the two charts (U_1, θ_1) and (U_2, θ_2) for S^1 which are defined by Figure 12.2.5. These charts are compatible with the submanifold structure of S^1 and have the property that

$$\theta_1 = \begin{cases} \theta_2 & \text{on one connected component of } U_1 \cap U_2 \\ \theta_2 - 2\pi & \text{on the other.} \end{cases}$$

Hence

$$\dot{\theta}_1 = \dot{\theta}_2 \quad \text{on} \quad TU_1 \cap TU_2 \ .$$

Thus we may define a smooth map, called the *angular velocity map*, by putting

$$\omega : TS^1 \longrightarrow R \quad \text{with} \quad \omega = \begin{cases} \dot{\theta}_1 & \text{on } TU_1 \\ \dot{\theta}_2 & \text{on } TU_2 \end{cases}$$

In addition to its obvious kinematic interpretation, the angular velocity has an important geometrical rôle as a component of the diffeomorphism, defined in Exercise 6.1.1,

$$\Phi : TS^1 \longrightarrow S^1 \times R$$

between the tangent bundle of the circle and the cylinder. It is set as an exercise to show that, in fact, this diffeomorphism satisfies

$$\Phi = (\tau_{S^1}, \omega) \tag{1}$$

The geometrical quantities needed for the study of the simple pendulum have now been defined. Before giving a formal discussion, however, we use physical intuition to guess at the possible types of orbit which can appear in the velocity phase space, as in Irwin(1980). Thus the orbits shown in Figure 12.2.7, shown in both chart and cylinder representation, can arise in the following ways:

(a) If placed with zero initial velocity at either the lowest or the highest point of the circle, the particle will stay there indefinitely. These two "equilibrium points" give rise to the two single-point orbits where the cylinder meets the y-axis. These two orbits are said to be "stable" and "unstable", respectively, for obvious reasons.

(b) If started with zero initial velocity at some point intermediate in height between the two equilibrium points, the particle will oscillate to and fro, rising indefinitely often to the initial height - first on one side of the origin, then on the other. These motions are represented on the cylinder by closed orbits which encircle the y-axis.

(c) If the particle is started with sufficiently large initial velocity it will pass through the highest point of the circle and then continue to make complete circuits. The corresponding orbits are the closed orbits which go right around the cylinder. There are two distinct families of these orbits because the particle may traverse the full circle either clockwise or anticlockwise.

It is left as an exercise to show that there exists an orbit having the unstable equilibrium point orbit as a limit point. A particle traversing this orbit approaches arbitrarily close to the equilibrium point without ever actually reaching it.

To complete our discussion of the motion of the simple pendulum, it remains to show how the formal theory from previous chapters leads to the sketches of (a) the vectorfields in Figure 12.2.6. and (b) the orbits in Figure 12.2.7.

As to the vectorfields, we first apply the theory of Section 8.2 to the particle constituting the bob of the pendulum. In terms of the x and y-coordinates the kinetic and potential energies are

$$T = \tfrac{1}{2}m(\dot{x}^2 + \dot{y}^2)$$

$$V = mgy.$$

After restricting these functions to TS^1 we get the Lagrangian function $L : TS^1 \to R$ by putting $L = T - V\circ\tau_{S^1}$. We will need to use both of the charts (U_1, θ_1) and (U_2, θ_2) for S^1 so let $i = 1$ or 2 and note that

$$\left.\begin{aligned} x &= \sin\circ\theta_i \\ y &= -\cos\circ\theta_i \end{aligned}\right\} \text{ on } U_i$$

and hence

$$L = \tfrac{1}{2}m\dot{\theta}_i^2 + mg\cos\circ\theta_i\circ\tau_{S^1} \quad \text{on} \quad U_i.$$

Thus, in terms of the local representative $\bar{\theta}_i = \theta_i \circ \tau_{S^1}\circ c$ of an integral curve $c\colon I \to TS^1$ with respect to the chart (U_i, θ_i), Lagrange's equation becomes

$$\bar{\theta}_i'' + g\sin\circ\bar{\theta}_i = \underline{0}. \qquad (2)$$

Hence, by the procedure given in Example 10.3.1, we find for $i = 1$ that the local vectorfield corresponding to this second-order differential equation is

$$Y_{T\theta_1} : T(-\pi,\pi) \longrightarrow T^2R$$
$$: (\phi,\omega) \longmapsto ((\phi,\omega),(\omega,-g\sin(\phi)))$$

This vectorfield is sketched on the left in Figure 12.2.6 and lies flat in the plane. The sketch for $Y_{T\theta_2}$ is similar. Now, by the construction in Section 11.1, the global vectorfield for the problem,

$$Y: TS^1 \longrightarrow T^2S^1,$$

is related to the above vectorfields by

$$Y = T(T\theta_i)^{-1} \circ Y_{T\theta_i} \circ T\theta_i \quad \text{on} \quad U_i \,.$$

Finally, the vectorfield sketched on the cylinder in Figure 12.2.6. is

$$Y_\Phi: S^1 \times R \longrightarrow T(S^1 \times R)$$

given by

$$Y_\Phi = T\phi \circ Y \circ \Phi^{-1}$$

where Φ is the diffeomorphism from TS^1 to the cylinder. It is possible to calculate an explicit formula for Y_Φ, although we shall not give the details here. Intuitively, this vectorfield is obtained by wrapping the local vectorfields $Y_{T\theta_1}$ and $Y_{T\theta_2}$ around the cylinder, at the same time allowing their arrows to poke out into R^3 while remaining tangent to the cylinder.

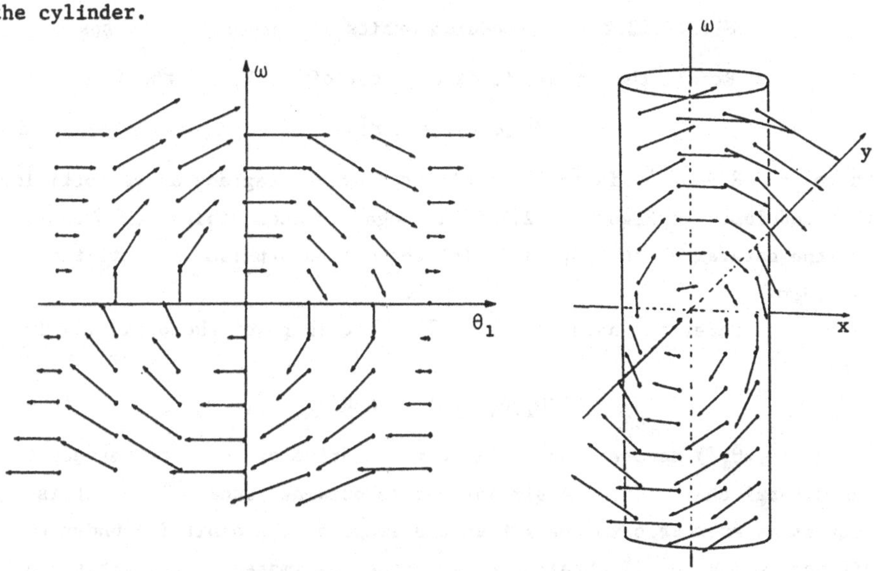

Figure 12.2.6. The vectorfield for the simple pendulum.

Turning now to the orbits, our aim is to formally justify the claims made earlier on the basis of physical intuition. In other words we are going to show how to derive mathematically the sort of orbits shown in Figure 12.2.7.

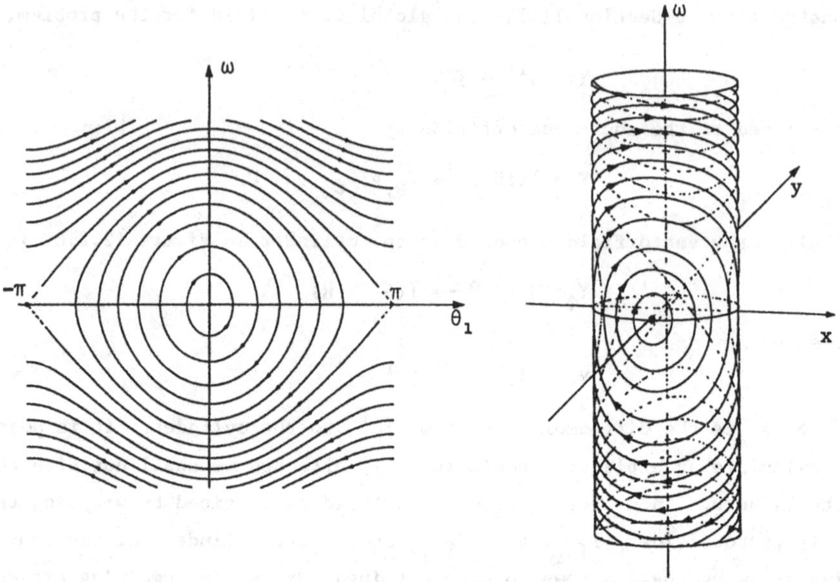

Figure 12.2.7. Pendulum orbits in velocity phase space.

Recall that an orbit is a subset of TS^1 of the form

$$\{c(t)\colon t \in R\} \tag{3}$$

where $c : R \to TS^1$ is an integral curve of the Lagrangian vectorfield of the pendulum. In Figure 12.2.7 the images of such orbits are shown under (a) the natural chart $T\theta_1$ and (b) the diffeomorphism Φ onto the cylinder.

Under the natural chart $T\theta_1$ the image of the orbit (3) is the set

$$\{(\bar{\theta}_1, \bar{\theta}_1{}')(t)\colon t \in R\} \subseteq TR ,$$

where $(\bar{\theta}_1, \bar{\theta}_1{}')$ denotes the local representative of c with respect to the natural chart $T\theta_1$. A similar set is obtained under $T\theta_2$. It is left as an exercises to check that the image of the orbit (3) under the diffeomorphism Φ is obtained by wrapping the images of the orbit under $T\theta_1$ and $T\theta_2$ around the cylinder.

To complete our discussion of the orbits we use energy considerations. Given $e \in R$, the set of all points in TS^1 at which the sum of the kinetic and potential energies assume the value e turns out to be a curve in TS^1 - called a *constant energy curve*. Since by Theorem 8.5.1 the total energy stays constant along an integral curve $c: R \to TS^1$, it follows that the orbits are subsets of the constant energy curves. The images of the constant energy curves under the charts $T\theta_1$ and $T\theta_2$ may be obtained by putting

$$\tfrac{1}{2}(\bar{\theta}_i')^2 - g \cos\circ\bar{\theta}_i = \underline{e} \qquad (4)$$

for $i = 1,2$ and then sketching the graph of $\bar{\theta}_i'$ against $\bar{\theta}_i$. This is, in fact, the way in which the curves shown in Figure 12.2.7 were obtained. Each orbit must lie inside a constant energy curve, although some constant energy curves contain more than one orbit. Since (4) implies that $e \geqslant -g$, the following cases exhaust the possibilities.

(a) $e = -g$. *Here the constant energy curve is a single point corresponding to the stable equilibrium point.*

(b) $-g < e < g$. *Here the constant energy curves are closed curves encircling the stable equilibrium point. Each curve is a single orbit.*

(c) $e = g$. *The constant energy curve contains the unstable equilibrium point. Removal of this point leaves two connected components, each of which is a single orbit.*

(d) $e > g$. *The constant energy curves encircle the cylinder. Each curve is a single orbit.*

These results will be verified in the exercises.

EXERCISES 12.2.

1. *Discuss the advantages of representing the orbits for the pendulum problem on a cylinder rather than in the plane via charts (see Irwin(1980), page 4).*

2. *Verify that the diffeomorphism* $\Phi: TS^1 \to S^1 \times R$ *defined in Exercise 6.1.1 can be written in the form (1) in the text.*

3. *Let* Φ *be as in the previous exercise and let* (U_i, θ_i) *be the chart for* S^1 *introduced in the text* $(i = 1,2)$. *Show that if* $c: R \longrightarrow TS^1$ *is an integral curve for the Lagrangian vectorfield of the pendulum then*

$$\Phi \circ c(t) = (\sin \circ \bar{\theta}_i, \cos \circ \bar{\theta}_i, \bar{\theta}_i')(t)$$

for all $t \in R$ *such that* $c(t) \in U_i$.

(This verifies the claim made in the text that the image of an orbit under Φ *is obtained by wrapping around the cylinder the images of that orbit under each of the charts* $T\theta_1$ *and* $T\theta_2$.*)*

4. *Verify from (4) in the text that the constant energy curves have the general shape claimed in the text for the various ranges of values of the parameter* e.

5. *Let* $c: R \longrightarrow TS^1$ *be an integral curve of the Lagrangian vectorfield of the pendulum. Note that if* $e > g$ *then the angular velocity,* $\omega \circ c$, *along the integral curve is bounded away from zero. Deduce that in this case the constant energy curve is a single orbit.*

6. *Let* $\bar{\theta} : R \longrightarrow R$ *be a differentiable function such that*

(a) $\bar{\theta}(t)$ *is bounded as* $t \longrightarrow \infty$

(b) $\lim_{t \to \infty} \bar{\theta}'(t)$ *exists.*

Show from the mean value theorem that

$$\lim_{t \to \infty} \bar{\theta}'(t) = 0.$$

7. *Figure 12.2.8 shows a constant energy curve for the pendulum in the case* $-g < e < g$, *which crosses the* θ_1*-axis at the points* $(-a,0)$ *and* $(a,0)$ *where* $0 < a < \pi$. *Let* $(\bar{\theta}_1, \bar{\theta}_1')$ *be the local representative of an integral curve and suppose that* $(\bar{\theta}_1, \bar{\theta}_1')(t)$ *lies on the constant energy curve and in the upper half plane when* $t = t_0$.

This exercise is to prove that at some later time $t > t_0$ *the point* $(\bar{\theta}_1, \bar{\theta}_1')(t)$ *passes into the lower half-plane. Suppose, on the contrary that it stays in the upper half-plane for all* $t > t_0$ *and then derive a contradication by showing, with the aid of Exercise 6, that*

(a) $\lim_{t \to \infty} \bar{\theta}_1(t)$ *exists*

(b) $\lim_{t \to \infty} \bar{\theta}_1'(t)$ *exists*

(c) $\lim_{t \to \infty} \bar{\theta}_1'(t) = 0$

(d) $\lim_{t \to \infty} \bar{\theta}_1''(t)$ *exists*

(e) $\lim_{t \to \infty} \bar{\theta}_1''(t) = 0$

(f) $\lim_{t \to \infty} \bar{\theta}_1''(t) \neq 0$

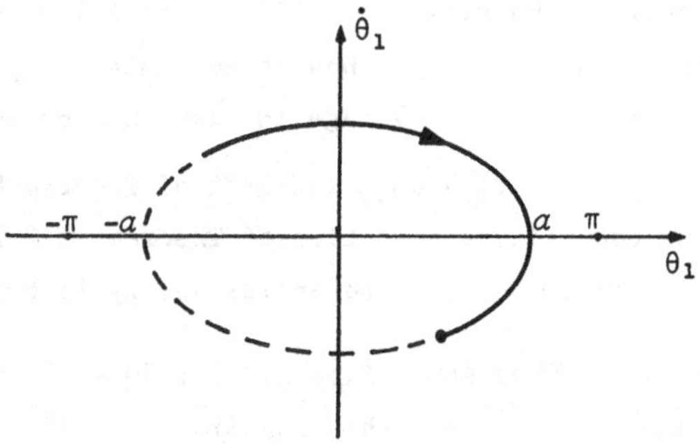

Figure 12.2.8.

8. *Deduce from the previous exercise that in the cases* $-g < e < g$ *each constant energy curve in the pendulum problem consists of a single orbit.*

9. *Suppose that* $Y\colon M \to TM$ *is a smooth vectorfield on the manifold* M *and let* $a \in M$ *with* $Y(a) = 0$ *(so that* a *is an equilibrium point of the vectorfield). Let* $c\colon R \to TM$ *be an integral curve of the vectorfield. Show that if* $c(t_0) \neq a$ *for some* $t_0 \in R$ *then* $c(t) \neq a$ *for all* $t \in R$.

10. *Prove the statements made in the text about the orbits of the pendulum in the case* $e = g$. *Use the results of Exercises 6 and 9.*

12.3. EXISTENCE OF LAGRANGIAN FLOWS

In this section the major result is a theorem which can be used to establish the completeness of many Lagrangian vectorfields arising from classical problems such as the simple pendulum, motion on a paraboloid (see Section 9.1) and the spherical pendulum (dealt with in Chapter 13). As noted in Section 12.1, completeness of a vectorfield implies the existence of a flow.

12.3.1. Theorem. *Let* Q *be a smooth k-dimensional submanifold of* R^n *and suppose* Q *is a closed subset of* R^n. *Suppose* $Y : TQ \to T^2Q$ *is a smooth second-order vectorfield on* TQ *and that*

$$c\colon (-\delta, \varepsilon) \to TQ$$

is a maximal integral curve for Y *with the property* $\|\pi_2 \circ c(t)\| \leq K$ *for some* K *and all* $t \in (-\delta, \varepsilon)$. *Then* $\delta = \varepsilon = \infty$.

Proof. The first part of the proof involves showing that for $t_n = \varepsilon - \frac{\varepsilon}{n}$ the sequence $c(t_n)$ has an accumulation point in TQ. The proof given by Chillingworth(1976) page 188 can then be modified to complete the proof. ■

The energy integral of Theorem 8.5.1 can be used to establish the boundedness condition of Theorem 12.3.1 in problems of classical mechanics where the potential energy is bounded below.

12.3.2. Theorem. *Suppose* $L : TQ \to R$ *is a smooth Lagrangian function where* Q *is a smooth submanifold of* R^{3n} *which is closed and let*

$$L = T - V\circ\tau_Q \quad \text{with} \quad T = \tfrac{1}{2}\dot{X}^T M\dot{X}$$

where $M = \mathrm{diag}(m_1, m_1, m_1, m_2, m_2, m_2, \ldots, m_n, m_n, m_n)$, $m_i > 0$.

If the potential energy map $V : Q \to R$ *is bounded below, then the Lagrangian vectorfield generated by* L *is complete.*

Proof. By Theorem 8.5.1, for each integral curve c and each $t \in I$,

$$(T + V\circ\tau_Q)(c(t)) = (T + V\circ\tau_Q)(c(0))$$
$$= e \quad \text{say.}$$

Thus, if d is a lower bound for V,

$$(\tfrac{1}{2}\dot{X}^T M\dot{X})(c(t)) \leqslant e - d$$

which gives the required result. ■

12.3.3. Corollary. *Let* L *and* Q *be as in Theorem 12.3.2. and suppose* Q *is a closed subset of* R^{3n}. *Then there exists a flow on* TQ *generated by the Lagrangian vectorfield corresponding to* L. ■

Finally we state a theorem in the context of a vectorfield on R^k which tells us that integral curves vary continuously with initial conditions. This property can readily be extended to vectorfields on manifolds.

12.3.4. Theorem. *Let* $c: (-\varepsilon,\varepsilon) \to U$, $d: (-\varepsilon,\varepsilon) \to U$ *where* U *is an open subset of* R^k, *be integral curves for the vectorfield* $F : U \to TU$. *Suppose* F *satisfies*

$$\|\pi_2\circ F(x) - \pi_2\circ F(y)\| \leqslant K\|x - y\|$$

for some $K > 0$ *and all* $x, y \in U$.

Then for each $t \in (-\varepsilon,\varepsilon)$

$$\|c(t) - d(t)\| \leq \|c(0) - d(0)\| e^{K|t|}.$$

Proof. See exercises. ■

EXERCISES 12.3.

1. *Explain why there exists a Lagrangian flow for the problems described in Example 12.2.2 and Section 9.1.*

2. *Prove Theorem 12.3.4 First note that* $c^{\cdot}(t) = F \circ c(t)$ *and so we can apply Theorem 1.5.4 (ii) to show that*

$$\int_0^t c' = c(t) - c(0)$$
$$= \int_0^t \pi_2 \circ F \circ c.$$

Now show that for $t \in (0,\varepsilon)$

$$\|c(t) - d(t)\| \leq \|c(0) - d(0)\| + \int_0^t K\, norm \circ (c-d)$$

where "norm" is given by norm: $R^k \to R$: $x \to \|x\|$. *Then apply Gronwall's inequality, Lemma 1.5.6, to get the required result. Finally show how to extend the result to all of* $(-\varepsilon,\varepsilon)$.

13. THE SPHERICAL PENDULUM

The spherical pendulum is typical of the sorts of problems which are traditionally studied in Lagrangian mechanics in that it has a configuration manifold which is 2-dimensional. In addition it has the very special simplifying property of being "integrable": in suitable charts Lagrange's equations "uncouple" leading, in effect, to a pair of 1-dimensional problems. A fairly extensive list of integrable problems is contained in Whittaker(1952) and a recent addition to this list is given in Gray et al(1982). Our discussion of the spherical pendulum will illustrate the way in which the preceding theory can be applied to such problems.

13.1. CIRCULAR ORBITS

A spherical pendulum consists of a particle of mass m constrained to move under gravity on a sphere, in the absence of friction. The constraint can be achieved, for example, by attaching the particle to one end of a light rod while keeping the other end fixed. The sphere on which the particle moves will be taken as

$$S^2 = \{(a,b,c) \in R^3 : a^2 + b^2 + c^2 = 1\}$$

which is a submanifold of R^3. The kinetic and potential energies are given in terms of the identity chart (x, y, z) on R^3 by

$$T = \tfrac{1}{2}m(\dot{x}^2 + \dot{y}^2 + \dot{z}^2)$$

$$V = gz .$$

The Lagrangian $L : TS^2 \to R$ is then the map $T - V \circ \tau_{S^2}$ restricted to the domain TS^2, so L is smooth.

By Theorem 11.2.3, Lagrange's equations, with respect to the various charts in any atlas for S^2, yield a second-order vectorfield on TS^2 - the Lagrangian vectorfield for the spherical pendulum. Since S^2 is compact, moreover, Theorem 12.3.2. shows that this vectorfield is complete. Hence it generates a flow on TS^2 by Theorem 12.1.9.

Thus there is a unique integral curve of the vectorfield at each point of TS^2 with R as domain. The integral of curves, furthermore, are self-consistent. Since there is no realistic way to sketch orbits on the 4-dimensional manifold TS^2, we shall have to be content to sketch the projections of these orbits on S^2. These projected orbits are traced out by the base integral curves.

The existence of the following families of orbits is almost obvious on the basis of physical intuition:

(a) *The equilibrium points at the north pole* $N = (0,0,1)$ *and the south pole* $S = (0,0,-1)$. *If placed at rest at either of these points, the particle stays there indefinitely.*

(b) *Orbits along each meridian, obtained by intersecting* S^2 *with a vertical plane through the poles. The motion of the particle along a meridian is the same as for a simple pendulum.*

(c) *Orbits around each parallel of latitude below the equator, obtained by intersecting* S^2 *with a plane perpendicular to the polar axis of the sphere. A particle moving in these orbits is called a "conical pendulum".*

These orbits are illustrated in Figure 13.1.1.

It is very easy to formally establish the existence of these orbits by using the equivalence we have established between Newton's second law and Lagrange's equations. To illustrate the method, we prove the following physically plausible result:

13.1.1. Proposition. *Let* $c: R \to TS^2$ *be an integral curve for the Lagrangian vectorfield of the spherical pendulum at the point* $(a,v) \in TS^2$ *where* a *is either of the two poles. For all* $t \in R$ *it then follows that*

(a) *if* $v = 0$ *then* $\tau_{S^2} \circ c(t) = a$

(b) *if* $v \neq 0$ *then* $\tau_{S^2} \circ c(t)$ *lies in the vertical plane through the poles containing the vector* $v \in R^3$.

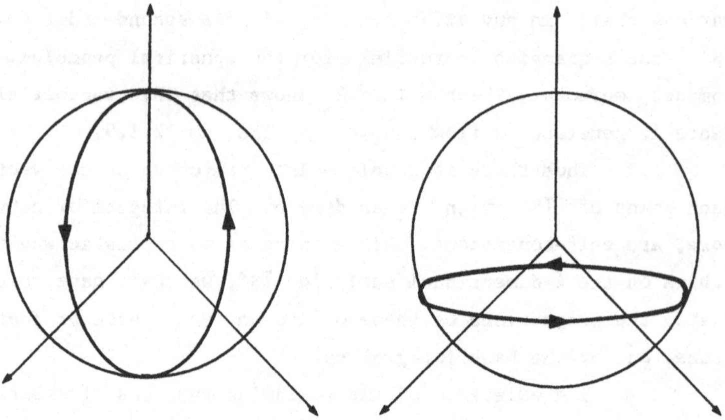

Figure 13.1.1. Motion on a meridian and on a parallel of latitude.

Proof. We leave the proof of part (a) as an exercise and prove part (b).

The theory for the simple pendulum given in Example 12.2.2 may be adapted to the meridian M as configuration manifold. There is then an integral curve $d: R \to TM$ at (a,v) for that problem. By Theorem 11.3.2, the reaction force R is orthogonal at each point to the circle M. It also lies in the plane of M and hence is orthogonal to S^2. Thus the integral curve d satisfies Newton's second law, as formulated in Section 8.2, for the spherical pendulum problem and hence, by Theorem 8.2.2 and Theorem 11.2.3, d is an integral curve for the Lagrangian vectorfield of the spherical pendulum at the point $(a,v) \in TS^2$. By uniqueness, $c = d$ and so $\tau_{S^2} \circ c$ maps into the meridian M. ■

EXERCISES 13.1.

1. *Prove part (a) of proposition 13.1.1.*

2. *Let* $d: R \to TS^2$ *be an integral curve for the Lagrangian vectorfield of the spherical pendulum. Show that if the base integral curve* $\tau_{S^2} \circ d$ *passes through a pole at some time* t_0 *then the base integral curve maps into a meridian.*

   (Hint: Put $c(t) = d(t + t_0)$ *and show* c *is an integral curve of*

of the vectorfield at some point (a,v) of the sort described in Proposition 13.1.1.)

3. *Read about the conical pendulum in Synge & Griffith (1959), pages 336-337, and then deduce the existence of corresponding integral curves for the Lagrangian vectorfield of the spherical pendulum.*

13.2. OTHER ORBITS, VIA CHARTS

The orbits of the spherical pendulum to be discussed in this section are not as obvious physically as those introduced in the previous section. To establish their existence, we shall choose charts for S^2 with respect to which Lagrange's equations "uncouple". Since orbits through the poles of S^2 have already been fully discussed in Section 13.1, moreover, we lose nothing by choosing charts whose domains exclude these points.

To define the desired charts put

$$U_1 = S^2 \backslash \{(a,b,c) \in S^2 : a \leqslant 0 \text{ and } b = 0\}$$

$$U_2 = S^2 \backslash \{(a,b,c) \in S^2 : a \geqslant 0 \text{ and } b = 0\}$$

and let the maps ϕ_1, ϕ_2, z be as shown in Figure 13.2.1 where

$$\phi_1 : U_1 \longrightarrow (-\pi,\pi)$$

$$\phi_2 : U_2 \longrightarrow (0,2\pi)$$

$$z : U_1 \cup U_2 \longrightarrow (-1,1) .$$

These charts are then $(U_i, (\phi_i, z|U_i))$ for $i = 1,2$.

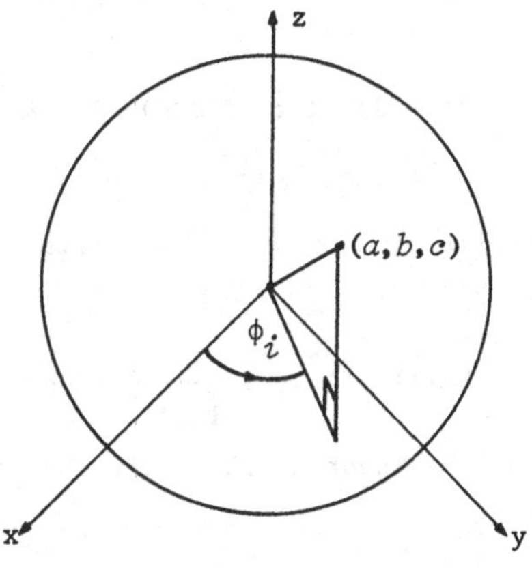

Figure 13.2.1. Charts for S^2 .

Note that $U_1 \cup U_2$, which is the domain of z, consists of the unit sphere in S^2 with both poles removed. On the other hand, $U_1 \cap U_2$ consists of two open hemispheres and

$$\phi_2 = \begin{cases} \phi_1 & \textit{on one hemisphere} \\ \phi_1 + 2\pi & \textit{on the other} \end{cases}$$

and hence we may define a smooth map $\omega : T(U_1 \cap U_2) \longrightarrow R$ by putting

$$\omega = \begin{cases} \dot{\phi}_1 & \text{on } TU_1 \\ \dot{\phi}_2 & \text{on } TU_2 \end{cases}$$

In this way we get a pair of maps, z and ω, which have a sort of global significance throughout the part of the manifold which now concerns us.

The differential equation introduced in the next proposition will play a key role in our discussion of the possible orbits for the spherical pendulum. As usual, we write

$$\bar{z} = z \circ \tau_{S^2} \circ c, \qquad \bar{\phi}_i = \phi_i \circ \tau_{S^2} \circ c$$

for the local representatives of an integral curve c.

13.2.1. Proposition. *Let* $c : R \longrightarrow TS^2$ *be an integral curve for the Lagrangian vectorfield of the spherical pendulum satisfying the condition that the corresponding base integral curve never passes through a pole. There are then numbers* e, $K \in R$, *depending on the initial value* $c(0)$, *such that*

$$(\bar{z}')^2 = 2g((1-\bar{z}^2)((e/g)-\bar{z}) - K^2/2g)$$

$$= f \circ \bar{z}, \text{ say.} \tag{1}$$

Proof. When expressed in terms of the charts the Lagrangian L is given on the domain U_i by

$$L = \tfrac{1}{2}m((1-z^2)\dot{\phi}_i^2 + \frac{\dot{z}^2}{1-z^2}) - mgz$$

for $i = 1,2$. Hence by Theorem 8.2.2 the integral curve satisfies the Lagrangian equations

$$((1-\bar{z}^2)\,\bar{\phi}_i')' = \underline{0} \tag{2}$$

$$\frac{\bar{z}''}{1-\bar{z}^2} + \bar{z}(\bar{\phi}_i')^2 + \frac{\bar{z}(\bar{z}')^2}{(1-\bar{z}^2)^2} + \underline{g} = \underline{0} \tag{3}$$

(It is perhaps reassuring to note that our assumptions ensure that $\bar{z}^2$ can never assume the value 1.) From (2) it follows that there is a constant $K \in \mathrm{R}$ such that

$$(1 - \bar{z}^2)\bar{\phi}_i' = \underline{K} \tag{4}$$

Since $\dot{\phi}_1$ and $\dot{\phi}_2$ are restrictions of the smooth map ω it follows, moreover, that K is the same for both choices of i. We may assume, furthermore, that $K \neq 0$ since otherwise (4) would imply that $\bar{\phi}_i' = \underline{0}$ and hence the base integral curve would be an orbit along a meridian, contrary to hypothesis.

Now since, by Theorem 8.5.1, the total energy of the system is conserved, there is an $e \in \mathrm{R}$ such that

$$\tfrac{1}{2}\left((1-\bar{z}^2)(\bar{\phi}_i')^2 + \frac{(\bar{z}')^2}{1-\bar{z}^2}\right) + g\bar{z} = \underline{e} \tag{5}$$

and, moreover, e is independent of i. Solving (4) for $\bar{\phi}_i'$ and then substituting in (5) gives, after some manipulation, the required equation (1). ■

13.2.2. Proposition. *The function* f *appearing in the previous proposition has a graph which looks like one of those in Figure 13.2.2. In particular,* f *must have three zeros* $\alpha_1, \alpha_2, \alpha_3$ *such that*

$$1 < \alpha_1 \leqslant \alpha_2 < 1 < \alpha_3 \; .$$

Both of the cases $\alpha_1 = \alpha_2$ *and* $\alpha_1 < \alpha_2$ *actually occur, for suitable choice of the initial value* c(0) *for the integral curve.*

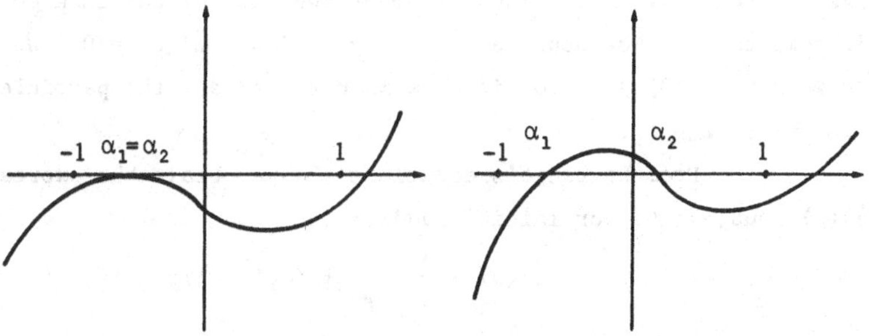

Figure 13.2.2.

Proof. Note first that, since $K \neq 0$, $f(-1) < 0$ and $f(1) < 0$. It is clear also that

$$\lim_{\alpha\to-\infty} f(\alpha) = -\infty \quad \text{and} \quad \lim_{\alpha\to\infty} f(\alpha) = \infty .$$

For motion to be at all possible, we must have $f\circ\bar{z} \geqslant \underline{0}$ and hence there must be at least one $\alpha \in (-1,1)$ such that $f(\alpha) \geqslant 0$. Hence the graph of f must be of one of the possible types shown.

To show that the case $\alpha_1 < \alpha_2$ can be realized by suitable choice of initial conditions note that from (1)

$$f(0) = 2e - K^2 .$$

But it is clear from (4) and (5) that we can keep K fixed while making e arbitrarily large by suitably choosing the initial conditions. This gives $f(0) > 0$ and hence $\alpha_1 < \alpha_2$.

Our argument that the case $\alpha_1 = \alpha_2$ can actually occur is indirect: from Section 13.1 there are orbits of the "conical pendulum" type; as shown later, these are not the sort of orbits obtained in the case $\alpha_1 < \alpha_2$; hence they must belong to the case $\alpha_1 = \alpha_2$. ■

The following result completes our discussion of the possible types of orbits for the spherical pendulum.

13.2.3. Proposition. *Suppose that the numbers α_1 and α_2 in the previous proposition satisfy the condition $\alpha_1 < \alpha_2$. There is then a family of orbits which oscillate between the parallels of latitude ("apsidal circles") on which the z-coordinate assumes the values α_1 and α_2 respectively, as in Figure 13.2.3.*

Proof. If $\bar{z}$ lies between α_1 and α_2 then $\bar{z}'$ cannot be zero since $f\circ\bar{z} > 0$ on (α_1, α_2). Thus $\bar{z}$ must increase or decrease monotonically in this range. Now suppose $\bar{z}(0) = \alpha_1$. Thus $\bar{z}'(0) = 0$. We need to show that $\bar{z}''(0) \neq 0$ for if this is not the case the particle would remain at $z = \alpha_1$.

From Lagrange's equations (3) we obtain the expression for $\bar{z}''(0)$ subject to our initial conditions

$$\bar{z}''(0) = \frac{\alpha_1 K^2}{\alpha_1{}^2 - 1} + (\alpha_1{}^2 - 1)g . \qquad (6)$$

Now $f(\alpha_1) = 0$ so from (1) we obtain

$$\frac{K^2}{\alpha_1{}^2-1} = 2g(\alpha_1 - \frac{e}{g})$$

and on substituting this into (6) we obtain

$$\bar{z}''(0) = g(2\alpha_1(\alpha_1 - \frac{e}{g}) + (\alpha_1{}^2-1)). \tag{7}$$

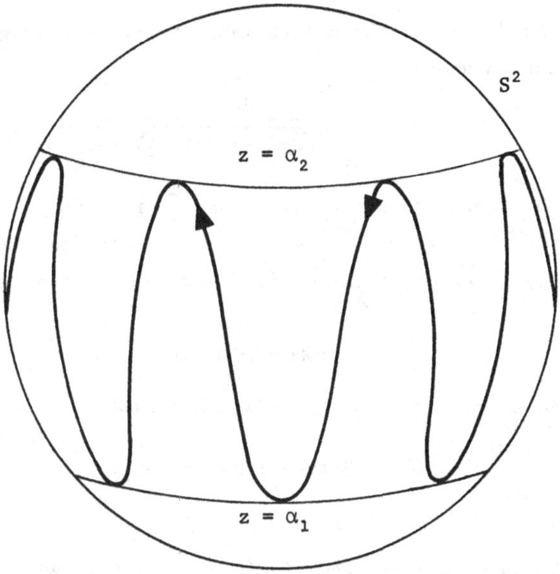

Figure 13.2.3. Oscillatory base integral curve.

It is easy to see that the right hand side of (7) is just $\frac{1}{2}f'(\alpha_1)$. But $f'(\alpha_1)$ cannot be zero otherwise α_1 would be a double root of f which would contradict our assumption that $\alpha_1 < \alpha_2 < 1 < \alpha_3$. Thus $\bar{z}''(0)$ is non-zero so that particle cannot remain at $z = \alpha_1$.

Now suppose we have initial conditions given by

$$\alpha_1 < \bar{z}(s) < \alpha_2, \quad \bar{z}'(s) > 0.$$

Suppose by way of contradiction that $\bar{z}$ does not reach $z = \alpha_2$ in finite time. Thus $\bar{z}(t) < \alpha_2$ whenever $t > s$. We deduced above that $\bar{z}$ increases monotonically in this range if $\bar{z}' > 0$ initially. So here we have

$$\bar{z}' = \sqrt{f\circ\bar{z}} \neq 0 \quad \text{on} \quad (\alpha_1,\alpha_2).$$

Thus

$$\int_s^t \frac{\bar{z}'}{\sqrt{f \circ \bar{z}}} = t - s$$

and so

$$\int_{\bar{z}(s)}^{\bar{z}(t)} \frac{dx}{\sqrt{2g(x-\alpha_1)(x-\alpha_2)(x-\alpha_3)}} = t - s \qquad (8)$$

The right hand side of (8) can be made arbitrarily large but the left hand side is bounded above by

$$\int_{\alpha_1}^{\alpha_2} \frac{dx}{\sqrt{2g(x-\alpha_1)(x-\alpha_2)(x-\alpha_3)}} \qquad (9)$$

which can be shown to be finite. Thus $\bar{z}$ reaches α_2 in finite time. The argument which shows $\bar{z}$ reaches α_1 is the same and the expression (9) gives the half-period of the motion between α_1' and α_2. ■

EXERCISE 13.2.

Assume $-1 < \alpha_1 \leqslant \alpha_2 < 1 < \alpha_3$ *and that*

$$\begin{aligned} f(\alpha) &= 2g(\alpha-\alpha_1)(\alpha-\alpha_2)(\alpha-\alpha_3) \\ &= 2g\left((\alpha^2-1)(\alpha - \frac{e}{g}) - \frac{K^2}{2g}\right). \end{aligned}$$

Use elementary algebra to show that $\alpha_1 + \alpha_2 < 0$ *and use this to show that the mean of the two "apsidal circles" shown in Figure 13.2.3 is below the* x-y *plane. What does this tell us about the "conical pendulum" base integral curves?*

14. RIGID BODY MOTION

In section 8.4 it was shown that the configuration set for the motion of a rigid rod was a 5-dimensional submanifold of R^6 *and furthermore that the reaction (or "internal") forces were orthogonal to this submanifold. Thus the problem could be treated as one of a single particle moving on a 5-dimensional submanifold of* R^6.

In this chapter the motion of a rigid body moving in R^3 *is considered and it is shown that the configuration space can be thought of as* $R^3 \times SO(3)$. *Thus this problem reduces essentially to the motion of a single particle on a 6-dimensional configuration manifold. Furthermore the reaction forces are shown to be orthogonal to the configuration manifold. The derivation of Lagrange's equations given in Section 8.3. thus applies to rigid body motion when the "external" forces are due to a conservative field of force.*

14.1. MOTION OF A LAMINA.

In section 8.4. we showed that Lagrange's equations were applicable to the problem of the motion of a rigid rod. There the "constraint condition" that the rod's length remained constant was sufficient to define the configuration manifold. By way of introduction to the problem of the rigid body we look at the motion on a plane of a "lamina" which we think of as being defined by three "point masses". Its configuration set will clearly be a subset of R^6 and we will again require that the distances between the particles remain constant. These three constraints, however, do not tell the whole story.

In Figure 14.1.1. we have two configurations which satisfy the same distance constraint conditions but if our lamina cannot "flip" (which is certainly the case because it is constrained to move in a plane) then these configurations cannot both be on the same configuration manifold.

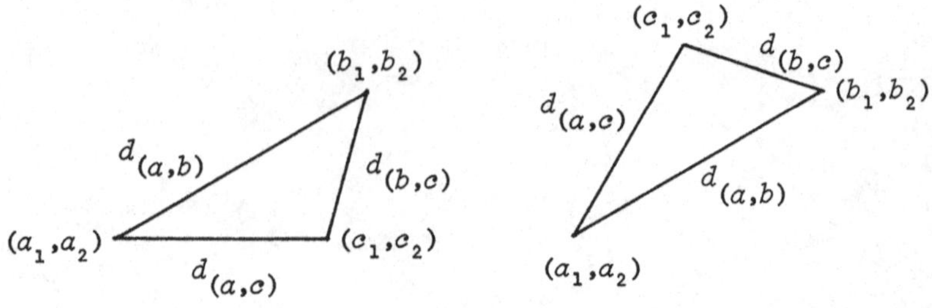

Figure 14.1.1.

Thus there must be another constraint which implies the preservation of the "orientation" of the lamina. This is obtained by the use of the vector or cross-product:

$$(b_1-a_1,\ b_2-a_2,\ 0) \times (b_1-c_1,\ b_2-c_2,\ 0)$$

which defines a vector normal to the plane in which the lamina moves. We will insist on this cross product also remaining constant.

The "distance preservation" constraint is $f(a_1,a_2,b_1,b_2,c_1,c_2) = 0$ where

$$f(a_1,a_2,b_1,b_2,c_1,c_2) = (u,v,w)$$

with

$$u = (b_1 - a_1)^2 + (b_2 - a_2)^2 - d^2_{(a,b)}$$

$$v = (c_1 - a_1)^2 + (c_2 - a_2)^2 - d^2_{(a,c)}$$

$$w = (b_1 - c_1)^2 + (b_2 - c_2)^2 - d^2_{(c,b)} .$$

Thus our configuration set Q satisfies $Q \subset f^{-1}(0)$. It turns out that $f^{-1}(0)$ is a three dimensional submanifold of R^6 which consists of the union of two disjoint open subsets, one of which is Q. The details are left as exercises.

An alternative way of describing the configuration manifold Q, which uses rotation matrices, is hinted at in the exercises. This alternative approach will be exploited in the next section where a direct description of the configuration manifold of a rigid body via constraints on distance and orientations becomes unweildy.

EXERCISES 14.1.

1. Read your solutions to Exercises 3.3.2. and 3.3.3. and then prove the statements made in the text concerning the sets Q and $f^{-1}(0)$.

2. Consider the following (distance preserving) transformation of a lamina which consists of an anticlockwise rotation through an angle θ about one of the vertices, which we place at the origin as in Figure 14.1.2.

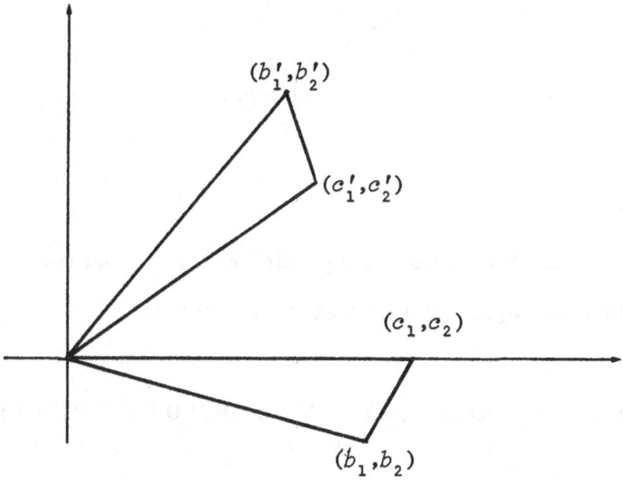

Figure 14.1.2.

(a) Find a matrix A such that

$$\begin{pmatrix} b_1' \\ b_2' \end{pmatrix} = A \begin{pmatrix} b_1 \\ b_2 \end{pmatrix}$$

$$\begin{pmatrix} c_1' \\ c_2' \end{pmatrix} = A \begin{pmatrix} c_1 \\ c_2 \end{pmatrix}$$

(b) Do these conditions determine the matrix A uniquely?

(c) Show that $\det(A) = 1$.

(d) Does this transformation preserve orientation?

3. Now consider the (distance preserving) transformation of a lamina which consists of a reflection about one of its sides, assumed to lie on the first axis, as in Figure 14.1.3.

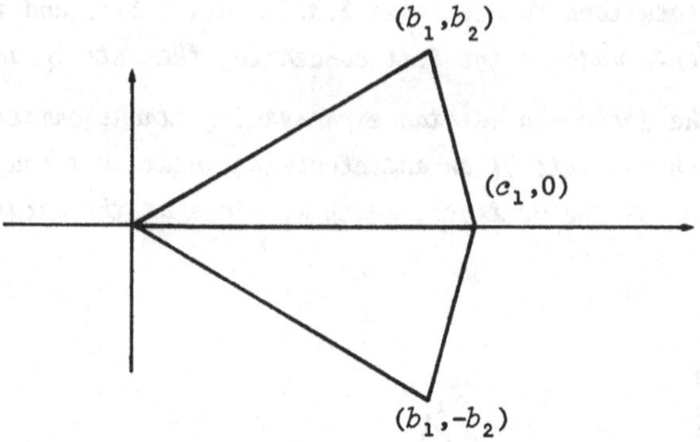

Figure 14.1.3.

(a) *Find the matrix of this transformation and its determinant.*

(b) *Does this transformation preserve orientation?*

14.2. THE CONFIGURATION MANIFOLD OF A RIGID BODY

Our mathematical model of a *rigid body* is a collection of n (≥ 4) particles in R^3, at least four of which are not coplanar. They are to be constrained in such a way that the distances between, and the relative orientation of, the particles remain constant throughout the motion. We regard these quantities as being preserved by means of rigid rods, of negligible mass, connecting each pair of particles.

To help us define the configuration set of the rigid body precisely, the following definition is introduced.

14.2.1. Definition. A *rigid motion* is a map $\rho : R^3 \to R^3$ which preserves

(a) distances between pairs of points in R^3,

(b) the orientation of triples of vectors in R^3. ∎

Algebraically, this definition means that the map ρ is to satisfy the conditions that

(a) $\|\rho(a)-\rho(b)\| = \|a - b\|$

(b) $\det(\rho(a),\rho(b),\rho(c)) = \det(a,b,c)$

for all a, b and c in R^3 where $\| \ \|$ is the usual norm on R^3.

The configuration set can now be defined as follows.

14.2.2. Definition. The *configuration set* Q of a rigid body is the set of n-tuples of vectors from R^3,

$$\{(\rho(a^1), \rho(a^2), \ldots\ldots, \rho(a^n)) : \rho \textit{ is a rigid motion}\},$$

where $a^1, a^2, \ldots\ldots, a^n$ are the respective initial positions in R^3 of the n particles constituting the rigid body. ∎

We now aim at showing that the configuration set is a submanifold of R^{3n}. Intuitively we can think of the configuration set of a rigid body as consisting of a combination of translations of one of its points together with rotations about axes passing through this point, as illustrated in Figure 14.2.1.

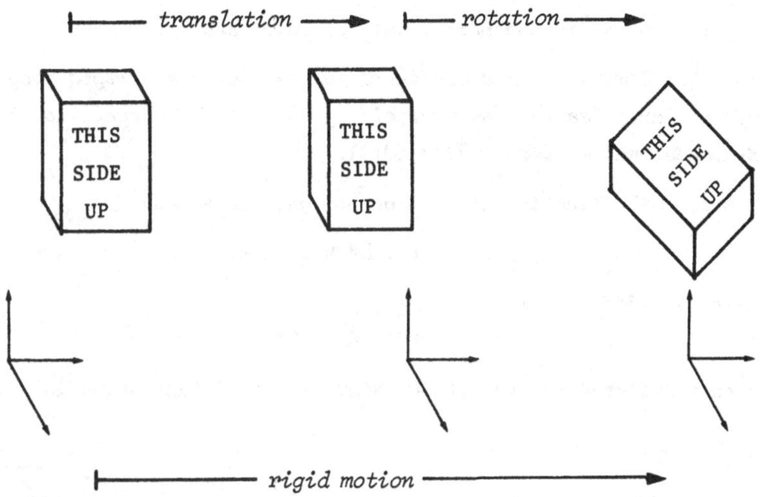

Figure 14.2.1. A rigid motion.

Thus it would seem that the configuration set may well look like

$$\{\textit{translations in } R^3\} \times \{\textit{rotations in } R^3\},$$

which can be represented as the manifold $R^3 \times SO(3)$. Our task then is to show that this is so. The key result is as follows:

14.2.3. Theorem. *A map* $\rho: R^3 \to R^3$ *is a rigid motion if and only if there is some* $c \in R^3$ *and* $A \in SO(3)$ *such that, for all* $a \in R^3$,

$$\rho(a) = c + Aa.$$

Proof. It is rather long and technical and appears at the end of this section. ■

In the following lemma, the element $A \in R^9$ is to be regarded as a 3×3 matrix.

14.2.4. Lemma. *Let* $L : R^3 \times R^{3\times 3} \to (R^3)^n$ *with*

$$L(c, A) = (c + Aa^1, c + Aa^2, \ldots, c + Aa^n)$$

where $n \geq 4$ *and* $a^1, a^2, \ldots, a^n$ *are the position vectors of points in* R^3 *not all lying in the same plane. The map* L *is then linear and one-to-one.*

Proof. This is left as an exercise. ■

By using the above map L we can easily derive the following theorem, which is the main result of this chapter.

14.2.5. Theorem. *The configuration set* Q *for a rigid body consisting of* n *particles is a submanifold of* R^{3n} *diffeomorphic to the 6-dimensional manifold* $R^3 \times SO(3)$.

Proof. The idea is that the one-to-one linear map

$$L : R^3 \times R^{3\times 3} \to R^{3n}$$

gives a diffeomorphism

$$L' : R^3 \times SO(3) \to Q$$

when restricted to the set $R^3 \times SO(3)$, as illustration in Figure 14.2.2.

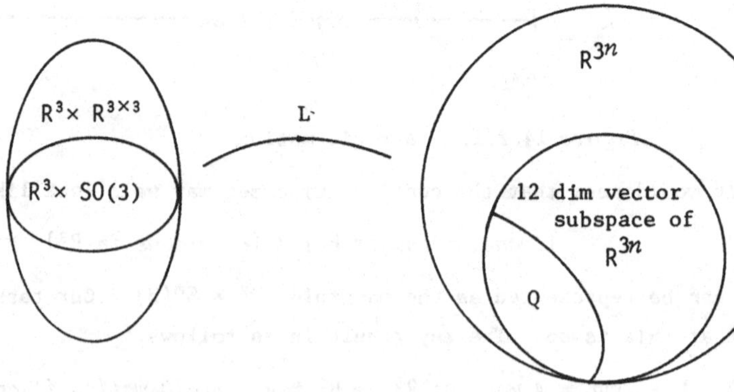

Figure 14.2.2.

The remaining details of the proof are left as an exercise. ■

We now go back and give a proof of Theorem 14.2.3, thereby completing this section.

Proof of Theorem 14.2.3. First, assume $\rho : R^3 \longrightarrow R^3$ is given by

$$\rho(a) = c + Aa$$

where $c \in R^3$ and $A \in SO(3)$. It follows from Lemmas 3.4.3. and 3.4.5. that ρ preserves distances between points and orientation of triples of vectors. Hence ρ is a rigid motion.

Conversely, assume $\rho : R^3 \longrightarrow R^3$ is a rigid motion. Let $L = \rho - \rho(0)$. It is easily checked that L preserves distances and orientations and that $L(0) = 0$. Thus, for each $a \in R^3$,

$$\|L(a)\| = \|a\|$$

and since, for all $a,b \in R^3$,

$$\|a-b\|^2 = \|a\|^2 - 2a.b + \|b\|^2$$

we have $$\|L(a)^2\|-2L(a)\cdot L(b) + \|L(b)\|^2 = \|a\|^2 - 2a.b + \|b\|^2$$

and so $$L(a)\cdot L(b) = a\cdot b\ .$$

Now let $u_i = L(e_i)$ for $i = 1,2,3$. Then for each i,j, $\|u_i\| = 1$ and $u_i\cdot u_j = \delta_{ij}$ and so $\{u_1,u_2,u_3\}$ is an orthonormal basis for R^3.

Now for each $i = 1,2,3$ and $a,b \in R^3$,

$$L(a).u_i = a\cdot e_i = a_i$$

$$L(b).u_i = b\cdot e_i = b_i$$

and so

$$(L(a + b) - L(a) - L(b))\cdot u_i = 0$$

which gives

$$L(a + b) = L(a) + L(b).$$

Similarly $L(\lambda a) = \lambda L(a)$ for each $\lambda \in R$. Thus L is linear and can be represented by some 3×3 matrix A whose columns are u_1,u_2,u_3. Thus

$$(A^TA)_{ij} = u_i\cdot u_j \quad \text{and so} \quad A^TA = I.$$

Now since L preserves orientations we must have, for three linearly independent vectors (a_1, a_2, a_3), (b_1, b_2, b_3) and (c_1, c_2, c_3),

$$\operatorname{sgn}\left(\left|\begin{matrix} A\begin{pmatrix}a_1\\a_2\\a_3\end{pmatrix} & A\begin{pmatrix}b_1\\b_2\\b_3\end{pmatrix} & A\begin{pmatrix}c_1\\c_2\\c_3\end{pmatrix}\end{matrix}\right|\right) = \operatorname{sgn}\left(\left|\begin{matrix}a_1 & b_1 & c_1\\a_2 & b_2 & c_2\\a_3 & b_3 & c_3\end{matrix}\right|\right)$$

Now letting

$$Z = \begin{pmatrix}a_1 & b_1 & c_1\\a_2 & b_2 & c_2\\a_3 & b_3 & c_3\end{pmatrix}$$

we have

$$\operatorname{sgn}(|AZ|) = \operatorname{sgn}(|A|)\ \operatorname{sgn}(|Z|) = \operatorname{sgn}(|Z|).$$

Thus $\operatorname{sgn}(|A|) = 1$ since $\operatorname{sgn}(|Z|) \neq 0$ and so $A \in SO(3)$. But

$$\rho(a) = \rho(0) + L(a) = \rho(0) + Aa$$

which shows that the map ρ has the desired form. ■

EXERCISES 14.2.

1. *Show that if* $\rho_1 : R^3 \to R^3$ *and* $\rho_2 : R^3 \to R^3$ *are rigid motions then so is the composite* $\rho_1 \circ \rho_2 : R^3 \to R^3$.

2. *Show that if an n-tuple of vectors* $(b^1, b^2, \ldots, b^n)$ *from* R^3 *belongs to the configuration set of a rigid body then the configuration set may be written as*

$$\{(\rho(b^1), \rho(b^2), \ldots, \rho(b^n)) : \rho \text{ is a rigid motion}\}.$$

3. *Prove Lemma 14.2.4. To show* L *is one-to-one, use the fact that if four elements of* R^3*, say* a^1, a^2, a^3, a^4*, are the position vectors of four non-coplanar points then the differences* $a^1-a^2, a^2-a^3, a^3-a^1$ *form a linearly independent set of vectors.*

4. *Complete the proof of Theorem 14.2.5. by using Lemma 4.3.4.*

14.3. ORTHOGONALITY OF REACTION FORCES

The methods employed here to study the reaction forces within a rigid body are basically the same as those used to study the corresponding problem for the rigid rod in Section 8.4. Here, however, the computational details are more complicated and it will be convenient to introduce some extra notation to describe the constraints between the n particles constituting the rigid body.

To this end we write a typical element $a \in R^{3n}$ in the form

$$a = (a^1, a^2, \ldots, a^n)$$

where, for $1 \leq i \leq n$, a^i is an element of R^3 which we write as

$$a^i = (a^i_1, a^i_2, a^i_3).$$

We think of a^i as the point occupied by the ith particle in R^3. For $1 \leq i < j \leq n$, furthermore, let d_{ij} denote the distance between the ith and j^{th} particles and let

$$f_{ij}: \quad R^{3n} \longrightarrow R \quad \text{with}$$

$$f_{ij}(a) = (a^i_1 - a^j_1)^2 + (a^i_2 - a^j_2)^2 + (a^i_3 - a^j_3)^2 - d_{ij}{}^2 \tag{1}$$

Finally, let $f: R^{3n} \longrightarrow R^{\frac{1}{2}n(n-1)}$ be the map whose respective real-valued components are the $\frac{1}{2}n(n-1)$ maps

$$\begin{array}{llll} f_{12}, & f_{13}, \ldots\ldots\ldots\ldots, f_{1n} \\ & f_{23}, \ldots\ldots\ldots\ldots, f_{2n} \\ & \quad \ddots \\ & \qquad\qquad f_{nn-1}\,. \end{array}$$

Since rigid motions preserve distance, it follows that the configuration manifold Q of the rigid body satisfies

$$Q \subseteq f^{-1}(0). \tag{2}$$

In the following geometrical lemma, notation is used for the components of $h \in R^{3n}$ which is analogous to that used for a above.

14.3.1. Lemma. *If* $(a,h) \in TQ$ *then, for* $1 \leq i \leq j \leq n$,

$$(a^i - a^j)\cdot h^i = (a^i - a^j)\cdot h^j \;.$$

Proof. Assume $(a,h) \in TQ$ so that $h = \gamma'(0)$ for some smooth parametrized curve γ in Q at a, by Definition 5.1.2. But by (2), for each t in an interval containing 0,

$$(f\circ\gamma)(t) = 0$$

and hence by the chain rule

$$Df(a)(h) = Df(\gamma(0))(\gamma'(0)) = D(f\circ\gamma)(t)(1) = 0.$$

In terms of the components of f this means that, for $1 \leqslant i < j \leqslant n$,

$$Df_{ij}(a)(h) = 0$$

The explicit form (1) of the map f_{ij} now gives the result. ■

As to the reaction forces between the particles constituting the rigid body, observe that the force exerted on the ith particle by the jth $(1 \leqslant i \leqslant n,\ 1 \leqslant j \leqslant n)$ will lie along the line joining these particles, as in Figure 14.3.1, and hence have the form

$$(a^i,\ \lambda_{ij}\ (a^j - a^i))$$

for some $\lambda_{ij} \in \mathsf{R}$. The reactions being equal in magnitude but opposite in direction, moreover, it follows that

$$\lambda_{ij} = \lambda_{ji} \tag{3}$$

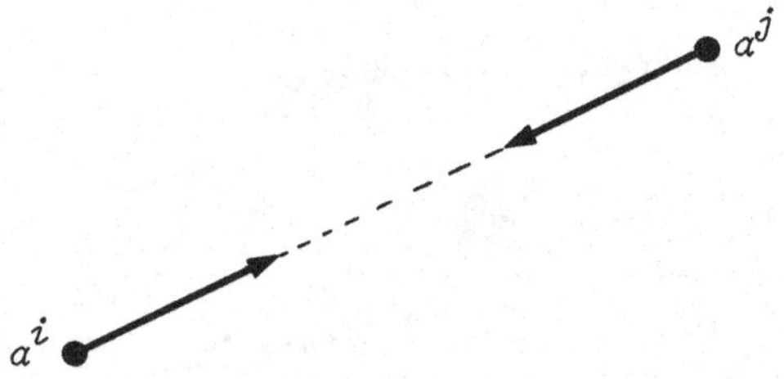

Figure 14.3.1. Equal and opposite reaction.

Now the total reaction force R^i on the ith particle will be the sum of such forces exerted by each of the other n-1 particles. Hence

$$R^i = \left(a^i,\ \sum_{\substack{j=1 \\ j \neq 1}}^{n} \lambda_{ij}\ (a^j - a^i)\right) \tag{4}$$

Hence, if we regard the rigid body as a single particle moving on the configuration manifold Q then the total reaction force acting on this particle at the point $a \in Q$ is given by

$$R = (a,\ (\pi_2(R^1),\ldots,\pi_2(R^n)\ )) \tag{5}$$

We can now state and prove the desired orthogonality result.

14.3.2. Theorem. *The reaction force* R *acting at any point of the configuration manifold* Q *is orthogonal to* Q *at the point.*

Proof. Let $(a,h) \in TQ$.

$$(a,R)\cdot(a,h) = \pi_2(R^1)\cdot h^1 + \dots + \pi_2(R^n)\cdot h^n \qquad \text{by (3)}$$

$$= \sum_{i=1}^{n} \sum_{\substack{j=1 \\ j\neq i}}^{n} \lambda_{ij}\,(a^j - a^i)\cdot h^i \qquad \text{by (4)}$$

$$= \sum_{\substack{i,j=1 \\ i<j}}^{n} \lambda_{ij}\,((a^j-a^i)\cdot h^i - (a^j-a^i)\cdot h^j) \qquad \text{by (3)}$$

$$= 0 \qquad \text{by Lemma 14.3.1.} \quad \blacksquare$$

REFERENCES

Abraham, R. & Marsden, J.E. (1978). Foundations of mechanics. (Second Edition). Reading, Mass.: Benjamin-Cummings.

Arnold, V.I. (1978). Mathematical methods of classical mechanics. New York: Springer Verlag.

Birkhoff, G.D. (1927). Dynamical systems. New York: American Mathematical Society.

Chillingworth, D. (1976). Differential topology with a view to applications. London: Pitman.

Courant, R. & John, J. (1974). Introduction to calculus and analysis volume 2. New York: Wiley

Dieudonne, J. (1960). Foundations of analysis. New York and London: Academic Press.

Dieudonne, J. (1972). Treatise on analysis. New York and London: Academic Press.

Goldstein, H. (1980). Classical mechanics. Reading, Mass.: Addison-Wesley.

Gray, A., Jones, A. & Rimmer, R. (1982). Motion under gravity on a paraboloid. J. Diff. Equations, 45, 168-181.

Gray, A. (1985). Motion under gravity on a saddle. J. Diff. Equations, 57, 248-257.

Guillemin, V. & Pollack, A. (1974). Differential topology. New Jersey: Prentice-Hall.

Hirsch, M.W. (1976). Differential topology. Graduate texts in Mathematics, Volume 33. New York: Springer.

Irwin, M.C. (1980). Smooth dynamical systems. New York and London: Academic Press.

Landau, L.D. & Lifshitz, E.M. (1960). Mechanics. Oxford: Permagon.

Lang, S. (1964). A second course in calculus. Reading, Mass.: Addison-Wesley.

Loomis, L.H. & Sternberg, S. (1968). Advanced Calculus. Reading, Mass.: Addison-Wesley.

Munroe, M.E. (1963). Modern multidimensional calculus. Reading, Mass.: Addison-Wesley.

Nering, E.D. (1964). Linear algebra and matrix theory. New York and London: Wiley.

Simmons, G.F. (1963). Introduction to topology and modern analysis. New York: McGraw-Hill.

Spivak, M. (1965). Calculus on manifolds. Reading, Mass.: Benjamin-Cummings.

Spivak, M. (1970). Differential geometry, Volume 1, Boston, Mass.: Publish or Perish.

Stern, R.J. (1983). Instantons and the topology of 4-manifolds. The Mathematical Intelligencer, 5, 39-44.

Synge, J.L. & Griffiths, B.A. (1959). Principles of Mechanics. New York: McGraw-Hill.

Whittaker, E.T. (1952). A treatise on the analytical dynamics of particles and rigid bodies. Cambridge: Cambridge University Press.

INDEX

SYMBOL TABLE

A^T	transpose of the matrix A
$\square$	empty set
$f \circ g$	composite of f with g, i.e. $(f \circ g)(x) = f(g(x))$
$\mathcal{L}(E,F)$	set of linear maps from E to F
N	natural numbers $\{1,2,3,\ldots\}$
R^n	euclidean n-space
R^+	strictly positive real numbers
$\|x\|$	norm of x in R^n
X	identity map on R^n
$Df(a)$	Fréchet derivative of f at a : 4
id_E	identity function on E : 5
$\underline{c}$	constant mapping : 5
T_aE	tangent space of E at a : 6
$\partial_i f$	partial Fréchet derivative of f with respect to the ith variable : 10
id_j	jth projection function : 10
$f'(a)$	jacobian matrix of f at a : 11
$f^{/i}$	slope of f in the ith direction : 11
$(f_1,f_2,..,f_m)$	components of vector valued function f : 12
$\int_a f$	indefinite integral of f from a : 13
S^n	unit sphere in R^{n+1} : 21
$\mathcal{C}^r$	r-times continuously differentiable : 24
Π	natural projection map : 29
$O(n)$	$n \times n$ orthogonal matrices : 35
$SO(n)$	$n \times n$ rotation matrices : 35
$[\gamma]_a$	tangent vector at a : 45
T_aM	tangent space of M at a : 46
TM	tangent bundle of M : 46
Tf	tangent map of f : 49
τ_M	natural projection from the tangent bundle : 60
T^2M	double tangent bundle of M : 62
$c^{\dot{}}(t)$	65

$\frac{\partial f}{\partial \phi_i}$	partial derivative of f with respect to ϕ_i : 68
$\frac{\partial f}{\partial \psi}$	partial derivative of f with respect to ψ : 70
$\dot{f}$	f dot : 72
$(\phi \circ \tau_Q, \dot{\phi})$	natural chart for TQ : 73
$f_{\phi\psi}$	local representative of f with respect to charts ϕ, ψ : 62
$\mathfrak{Diff}^\infty(M)$	set of all $\mathfrak{C}^\infty$ diffeomorphisms from M to M : 122
Λ	flow : 123